Gerardo Barbieri

AUTOMOTIVE SUSPENSION
Setting and fine-tuning of elastokinematics
to improve vehicle dynamics

Le Penseur Publisher

Automotive Suspension
Setting and fine-tuning of elastokinematics to improve vehicle dynamics

Gerardo Barbieri

ISBN 978-88-95315-94-2

Original Edition (Italian language), July 2022

First Edition (English language), April 2023

Printed in Italy, April 2023

Cover image: 3d vehicle model with suspension and steering © Cristian Li Voi

Author contact: gerrybarbieri@libero.it

© 2022 Le Penseur

   Via Montecalvario 40/3 - 85050 Brienza (Potenza) – ITALY

   http://www.lepenseur.it

All right reserved.

No part of this book may be reprinted or reproduced or utilized in any form or by any electronic, mechanical, or other means, now known or hereafter invented, including photocopying and recording, or in any information storage or retrieval system, without permission in writing from the publisher.

Product or corporate names may be trademarks or registered trademarks, and are used only for identification and explanation without intent to infringe.

Every effort has been made to contact and acknowledge copyright owners. If any material has been included without permission, the publisher offers its apologies. The publisher would be pleased to have any errors or omissions brought to its attention so that corrections may be published at later printing (please contact **info@lepenseur.it**).

The author and publisher of this book have used their best efforts in preparing the material in this book. The author and publisher shall not be liable in any event for any damages, including incidental or consequential damages, lost profits, or otherwise in connection with any text and methods explained in this book.

Index

FOREWORD — 9

CHAPTER 1. VEHICLE DYNAMICS – Maneuvers and goals — 11

1.1. The vehicle from a dynamic point of view — 11

1.2. What do we ask to the vehicle? — 11
    1.2.1. Straight line driving — 12
    1.2.2. Cornering — 12
    1.2.3. Slalom — 14

1.3. What does the dynamics vehicle behavior depend upon? — 16

CHAPTER 2. TIRE ROAD INTERACTION – Notes — 19

2.1. The quadricycle — 19

2.2. The slip angle — 20

2.3. The brush model — 23

2.4. *Camber* effects on the tire characteristic — 30

2.5. Other effects on the tire characteristic — 33
    2.5.1. The coefficient of adherence — 33
    2.5.2. The temperature and the inflation pressure — 37
    2.5.3. The vertical load — 39

2.6. The load transfer and the slip angle — 39

2.7. Summary — 42

CHAPTER 3. THE MOTIONS OF THE VEHICLE — 43

3.1. General considerations — 43

3.2. Sprung and unsprung masses — 43

3.3. The pitch and rebound motions — 45

3.4. The bump stiffness — 46

3.5. The sizing of the springs — 48

3.6. The roll and yaw motions — 52

3.7. The roll stiffness — 54

3.8. The sizing of anti-roll bar — 55

3.9. The load transfer in stationary condition — 58

3.10. The distribution of load transfers in stationary conditions — 61

3.11. The roll angle — 67

3.12. The sizing of the roll stiffness — 67

3.13. The sizing of the shock absorbers — 71

3.14. The dynamic influence of the shock absorbers — 75

3.15. Summary — 76

CHAPTER 4. LATERAL DYNAMICS – Use of analytical models — 79

4.1. The analytical approach — 79

4.2. The evaluation of the slip angles — 80

4.3. The acceleration and the evaluation of center of gravity velocity — 82

4.4. The single-track model — 84

4.5. The sideslip angle gradient — 86

4.6. The understeer gradient — 89

4.7. No stationary conditions — 92

4.8. Summary — 98

CHAPTER 5. SUSPENSION – Architectures and schematizations — 99

5.1. The schematic of the suspensions — **99**

    5.1.1. The elastic bushings — 102

5.2. Front suspensions — **105**

    5.2.1. Double wishbone — 105

    5.2.2. McPherson — 108

5.3. Rear suspensions — **111**

    5.3.1. Twist beam axle — 112

    5.3.2. Rear Double wishbone and rear McPherson — 114

    5.3.3. Bilink — 115

    5.3.4. Multilink — 116

    5.3.5. Trailing arm and semi-trailing arm — 118

CHAPTER 6. SUSPENSION – Set-up and operation — 123

6.1. General description — **123**

6.2. The characteristic angles of the wheels and the geometric quantities of the suspension — **124**

6.3. The *Camber* — **126**

6.4. The *Toe* — **130**

6.5. The effect of the *toe* on the lateral characteristic of the tire — **138**

6.6. *Toe* variation under longitudinal load — **142**

6.7. The kinematics of the suspension — **143**

    6.7.1. The roll center — 149

    6.7.2. *Camber* variation in bump travel — 152

    6.7.3. *Camber* variation when cornering — 153

    6.7.4. Alteration of the kinematics in the roll transient due to the shock absorbers — 159

6.8. Longitudinal kinematics — **161**

    6.8.1. Calculation of the ICSV position — 167

CHAPTER 7. THE STEERING SYSTEM — 171

7.1. The *Caster* — 171

7.2. The *Caster trail* — 176

7.3. The longitudinal arm at wheel center — 179

7.4. The King Pin Inclination — 179

7.5. The King Pin Offset — 180

7.6. The scrub radius — 183

7.7. The steering of the wheels — 186

7.8. The steering mechanism — 192

CHAPTER 8. VEHICLE DYNAMICS – Performance evaluation — 197

8.1. The evaluation tools — 197

8.2. Steering pad — 199

    8.2.1. The understeer curve — 199

    8.2.2. The sideslip angle — 204

    8.2.3. The roll gradient — 205

8.3. The response of the vehicle in stationary — 206

8.4. Step steer at constant speed — 208

    8.4.1. Yaw velocity and sideslip angle — 212

    8.4.2. Lateral acceleration and sideslip angle — 215

    8.4.3. The relaxation length — 221

    8.4.4. The roll angle — 222

    8.4.5. Considerations on different behaviors — 223

8.5. Summary — 225

CHAPTER 9. VEHICLE DYNAMICS – Achievement of the targets — 227

9.1. Straight line driving — 227

9.2. Cornering — 233

9.3. Slalom — 243

Analytical index	251
List of main symbols and abbreviations	259
Bibliography	265

IMPORTANT WARNING – Disclaimer

This book should not be considered a guide to modifying, designing or building an automobile or its suspension. Who uses it as such does it at their own risk. Testing vehicles on the road can be dangerous. The author and publisher are not responsible for any damages, direct or indirect, caused by the application of any information contained in this book.

The illustrations on pages 100, 101, 102, 111, 112 and 115 are based on images taken from the web. Being images published on numerous websites, it was not possible to identify the owners/authors. Should the owners/authors of the respective images be notified, the publisher will take care to mention them in future editions.

Foreword

This volume, as explained by the title, deals with the setting and fine-tuning of the elastokinematics of the suspension.

It consists of three main parts which respectively define some operating maneuvers, the knowledge to be acquired to interpret them from a physical point of view and the actions to be carried out to improve the dynamic behavior of the vehicle.

The first part, consisting of the first chapter, describes and schematizes some of the most frequent maneuvers in car driving and the related expected behaviors. Furthermore, the subsystems and the physical characteristics of the vehicle (arrangement of the masses and dimensions) that influence these behaviors are highlighted. The exercise maneuvers described are the straight line driving, the cornering (corner entry, transitory, steady state, corner exit) and the slalom.

The second part, which includes chapters 2 to 8, analyzes in detail the influence of the variability of subsystems and physical characteristics on the resulting dynamic performance of the vehicle.

Finally, the third part, consisting of the ninth chapter, proposes the exercise maneuvers described in the first chapter and reinterprets them using the knowledge acquired in the second part of the text.

The proposed aim is to identify which changes to make to the subsystems (suspension and steering) and to the dimensional characteristics of the vehicle to achieve the desired dynamic behavior.

In light of how the text has been structured, a consequential reading of the various chapters is necessary for its full understanding.

Gerardo Barbieri

1. VEHICLE DYNAMICS
Maneuvers and goals

1.1. The vehicle from a dynamic point of view

The vehicle is a physical system, maneuverable by using controls, that returns a dynamic behavior according to the contents of which it is made up.
The controls to handle it are:

1. the steering wheel;
2. the brakes;
3. the gas pedal;
4. the gearbox.

The contents that influence the dynamic behavior are:

1. the sizes and arrangement of masses
2. the steering system
3. the suspensions
4. the tires
5. the braking system

Maneuvering a vehicle means to give controls to obtain the desired behavior. A vehicle is dynamically performing when the behavior obtained, through the administration of controls, is always aligned with what is desired.

1.2. What do we ask to the vehicle?

The main driver ambition is to direct the vehicle according to the desired trajectories under all driving conditions. The achievement of this aim depends on the ability of the designer who must correctly size the above contents so that, under imposed controls, the desired behavior is obtained. Good sizing of the suspension and of the steering system allows to make the most of the adherence potential of the tires and to maneuver the vehicle with greater ease and mastery. To achieve this it is necessary that the contents are well integrated and that allow the vehicle to return, during its use, the right information about the road surface adherence conditions so as to establish the correct feeling with the user. In order to define a methodology for sizing the contents of the vehicle, we examine three maneuvers which, combined, reproduce most of the operating conditions. They are: *Straight line driving*; *Cornering*; *Slalom*.
Understanding the dynamics of these drive situations requires the knowledge of all the topics developed in the following chapters and will serve to give practical *feedback* to what has been learned in reading. We only briefly describe the above maneuvers, post-

poning the detailed explanation to the last chapter in which they will be re-proposed using the knowledge gained from understanding the text. The goal is to gain maximum confidence about the actions to be taken to improve the vehicle behavior in different situations.

1.2.1. Straight line driving

The different straight-line driving are as follows:

1. Straight line driving at constant speed (zero longitudinal acceleration);
2. Straight line driving at increasing speed (longitudinal acceleration, due to engine torque, concord to driving direction);
3. Straight line driving at decreasing speed (longitudinal acceleration, due to braking, opposite to driving direction).

In all three conditions the driver expects that, at steering wheel aligned, the vehicle maintains the straight direction even in the presence of random external disturbances that cannot be controlled by the driver.

1.2.2. Cornering

In this maneuver the vehicle moves along a curvilinear trajectory with variable radius. Regardless of the type of trajectory chosen by the driver, in order to individuate the factors on which depends the vehicle behavior, it was decided to divide the maneuver into the following four phases which, in reality, are not separate but overlapped.

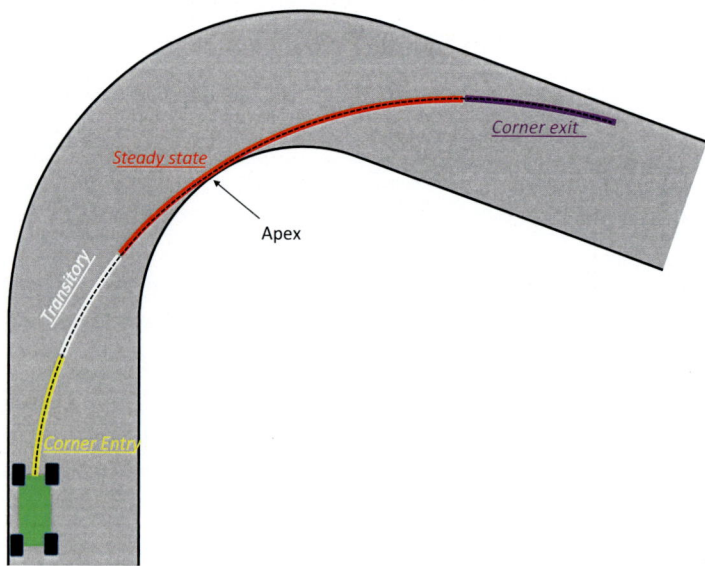

Figure 1 – Fictitious division of the curve maneuver into four phases which are normally superimposed. The figure highlights the apex which is usually located in the center of the curve.

Corner entry

It represents the time interval in which, starting from a straight driving condition, the driver applies the steering wheel angle they deem necessary to direct the vehicle to the apex of the curve, when driving on the track, or to follow the trajectory imposed on the roadway when driving on the road.

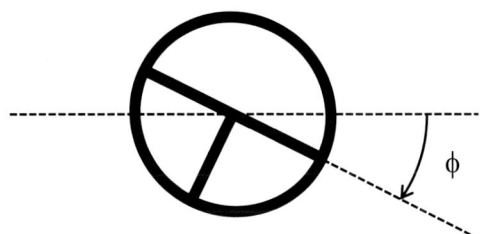

Figure 2 – Steering wheel angle imposed in the first part of cornering.

Transitory

It represents the time interval necessary for the vehicle to stabilize. This interval ends when roll angle reaches a constant value and the suspensions assume a certain configuration.

Normally during the transitory, the driver corrects the steering wheel angle to maintain trajectory which otherwise would tend to vary due the variation of orientation of the wheels caused by the suspension travel. The correction generates an additional *roll angle* and an additional delta of suspension travel. For optimal and intuitive driving, this correction must be concord with initial steering wheel angle ϕ. The need to counter-steer to keep the vehicle on trajectory indicates poor dynamics behavior because, in general, when it is required to apply a control contrary to the one just imposed, the user loses confidence in his ability to maneuver the vehicle and therefore, the driving feeling is lowered.

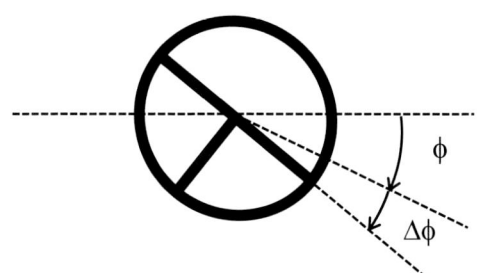

Figure 3 – Necessary correction to keep the vehicle on trajectory.

Steady state

It represents the interval of time in which the vehicle turns in stationary condition with constant steering wheel angle.

Corner exit
It represents the phase in which the user releases the steering wheel and accelerates, if the curve has been approached correctly. If, on the other hand, the trajectory has not been set well by the user, it represents the phase in which the user must increase the steering wheel angle to stay on track.

Regardless of the various driving techniques, each of the four phases of cornering maneuver can be carried out in the following three conditions.

- **Under acceleration.** This is when a driving torque is applied to the driving axle which causes a progressive increase in vehicle speed.
- **Under braking.** This is when a brake torque is applied to the four wheels with a consequent decrease in vehicle speed.
- **Under stationary condition.** This is when the driving torque is balanced with all the resistances (aerodynamics, rolling, etc.)

1.2.3. Slalom

The slalom is a very challenging maneuver that is usually performed in car races and in emergency situations. It consists of a sudden change in direction in order to make a double obstacle passing and then re-enter the trajectory, as shown in the following figure.

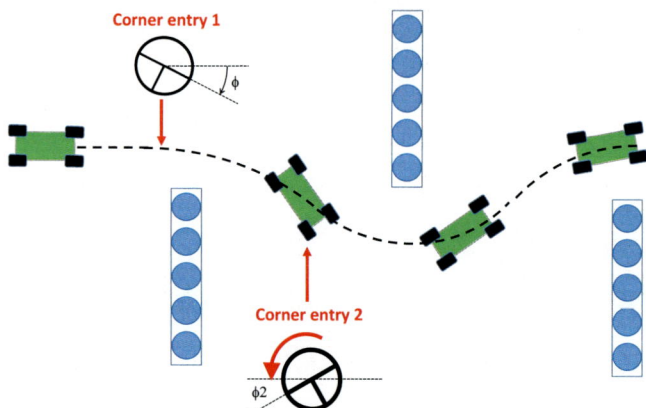

Figure 4 – Slalom schematization in which the obstacles consist of three rows of cones.

The first three phases of the slalom are the same of the cornering maneuver with the difference that in this case the change in magnitudes of the variables involved is much faster. We shall describe briefly the phases:

Corner entry 1
It represents the time interval in which, starting from a straight running condition, the driver applies the steering wheel angle ϕ they deem necessary to change direction. This phase has no difference with the corner entry maneuver described above.

Transitory 1
It represents the time interval in which the vehicle completes the roll travel and the user corrects the steering wheel angle to keep the trajectory. When the obstacle passing maneuver is very fast, the roll travel is completed only partially, and the correction of the steering wheel is minimal.

Steady state 1
It represents the interval of time in which the car turns with the stabilized quantities. This phase does not exist when the obstacle passing maneuver is very fast.

Corner entry 2
It represents the time interval in which the user applies the steering wheel angle $\phi 2$, in the opposite direction to the initial entry, in order to pass the next obstacle and return in the initial straight trajectory. This maneuver is delicate because the steering imposed by the user is much faster than the response of the car that is still on the previous roll phase. Consequently, the vehicle must turn with the outer wheels still extended, as shown in the following figure.

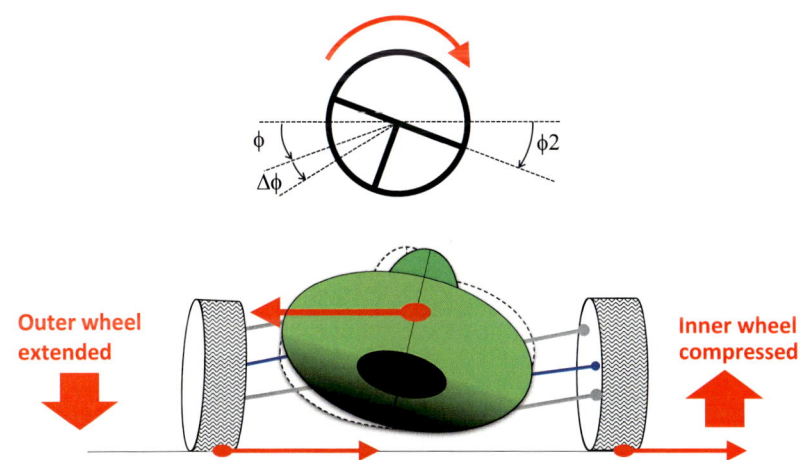

Figure 5 – Frontal view of vehicle at the beginning of corner entry 2 with the details of configuration of inner wheel (compressed) and of outer wheel (extended).

Transitory 2
The interval of time in which the total roll travel is completed with the consequent correction on steering wheel angle. This phase is different from transitory 1 for the greater roll excursion due to the fact that, when the steering wheel angle is applied, the car is rolled to the opposite side.

Re-enter in trajectory
The time interval in which the user applies the steering wheel angle that allows them to re-enter the trajectory. There is no particular criticality at this stage.

1.3. What does the dynamics vehicle behavior depend upon?

The dynamics performance of the vehicle depends on the contents with which it is made up. In order to have a certain dynamic response it is necessary:

1. Arrange the masses appropriately;
2. Define the correct steering ratio (Steering wheel angle/Wheel angle);
3. Identify the right front and rear suspension architecture;
4. Define the positions of the suspension components, in order to control wheels movements relative to the vehicle body;
5. Correctly size the springs and the anti-roll bars;
6. Correctly size the shock absorbers;
7. Identify the right size and type of tires.

We are leaving aside the aerodynamic influence in this publication. With regard to that, the reader shall refer to specialized texts as its complexity deserves a dedicated study. In order to give a preview of the problems, we briefly develop these points by postponing the detailed description to the following chapters.

Point 1

The vertical arrangement of the masses defines the height of the center of gravity. In order to improve the dynamics performances, it is preferable to arrange the masses so as to lower the center of gravity as much as possible and increase the wheel track. In fact, if H_G is the height of the center of gravity and t the wheel track of the vehicle, it can be seen that if H_G/t increases, with the same centrifugal force Fc, the $\varDelta Fz$ load transfer increases, bringing the vehicle closer to roll over and the external tires to the crisis.

$$\varDelta Fz = Fc \, \frac{H_G}{t}$$

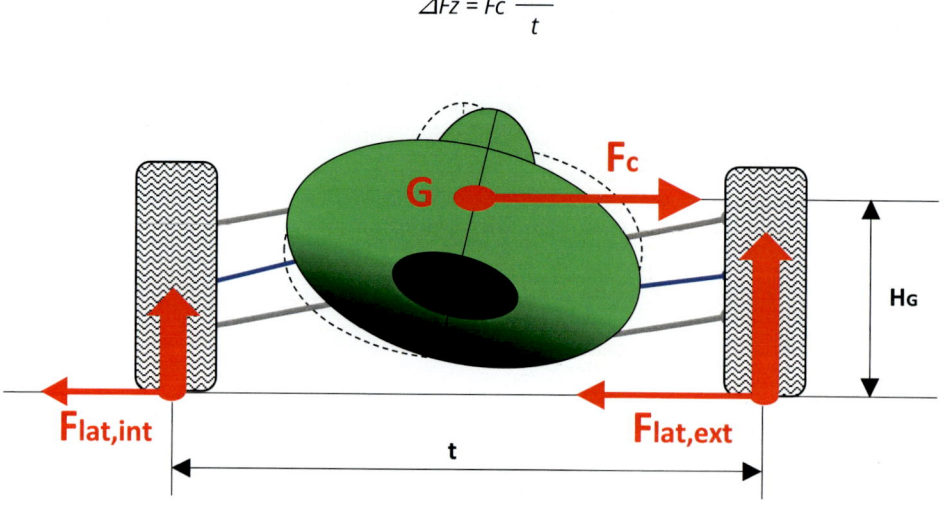

Figure 6 – Front view of the vehicle with the height of the center of gravity (H_G) affecting the moment of the lateral force.

The longitudinal position of center of gravity affects the slip angle, assumed by each axle, which in turn affects the understeer and oversteer of the vehicle.

Point 2
The steering ratio defines the steering wheel angle to turn according to a certain radius of curvature. It depends on the pinion and rack ratio, within the rack housing (steering box), and on the length of the steering lever. The more the steering ratio decreases, the more "direct" the car is.

Point 3
The suspension architecture defines the limits of the possible movements of the wheel with respect to the vehicle body. A McPherson type architecture, for example, has a limited *camber* gain compared to a Double wishbone architecture which, having a greater number of components, allows a greater range of motion.

Point 4
Once the suspension architecture that allows defined performance has been chosen, the components (arms, joints, etc.) must be positioned appropriately to have the desired relative movement between the wheel and vehicle body. One of the main aims to pursue is to determine the orientation that the wheels assume during the suspension travel and under the various operating loads. The orientation assumed by the wheels affects both the trajectory of the vehicle with steering wheel angle imposed and the interaction forces between the tire and the road.

Point 5
The sizing of the springs and anti-roll bars is important to achieve both the desired roll angle due to centrifugal force and the distribution of the slip angles on the two axles, and therefore for the understeer and oversteer tendency.

Point 6
The sizing of the dampers affects the time that the vehicle takes to stabilize during various operating maneuvers. The roll angle that the vehicle assumes when cornering, before stabilizing at the stationary value, undergoes an unwanted oscillation. It is the duty of the shock absorbers to dampen this vibration to reach the stationary value in the shortest time possible.

Point 7
The size of the tires is decided based on the weight of the vehicle. This influences the tendency to understeer when different sections are employed between the front and rear axles. Generally, cars that have the mass shifted towards the rear axle employ greater sections on the rear.

2. TIRE ROAD INTERACTION
Notes

2.1. The quadricycle

In the study of lateral dynamics, we can schematize the vehicle as a rigid body moving with planar motion on the plane of the road thanks to the rolling of the four wheels, anchored to the frame through bearings. To allow the vehicle to be steered, the front wheels can also rotate around an axis. In this chapter we will assume this axis to be vertical in order not to add excessive complications to the study of tire-road interaction. This schematization, which does not always take into account dynamics of roll and pitch, helps us to interpret much more easily the lateral behavior of the vehicle and is called *equivalent quadricycle*.

We shall apply a *steering wheel angle* to a vehicle which initially drives straight and has the ability to steer as shown in the following figure. According to the plane motion schematization described above, the vehicle (in steady state conditions) rotates around a point called instant center of rotation (I.C.), identifiable by joining the perpendiculars to the velocities of the four wheels centers. The latter should be parallel to the longitudinal tire planes, since they are constrained to follow the rolling, as shown in the following figure.

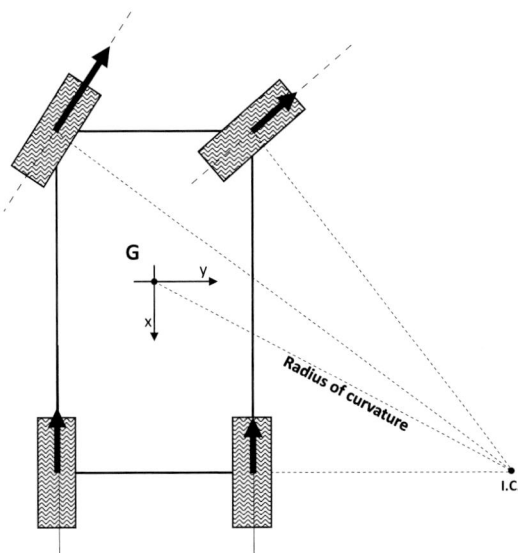

Figure 1 – Quadricycle with detail of wheel centers velocity, in steady state condition.

Automotive suspension

Thus, knowing the steer angles, derived from the steer mechanism, it is easy to identify the I.C. and the subsequent other quantities. Unfortunately, it is not so easy because the tires flexibility causes the wheel centers to have velocities, and therefore displacements not parallel to the corresponding longitudinal planes. Therefore, the question becomes more complicated because, in order to identify the I.C., it is necessary to determine the inclinations of wheel center velocity with respect to the longitudinal planes of the corresponding tires, and carry out the construction of the following figure.

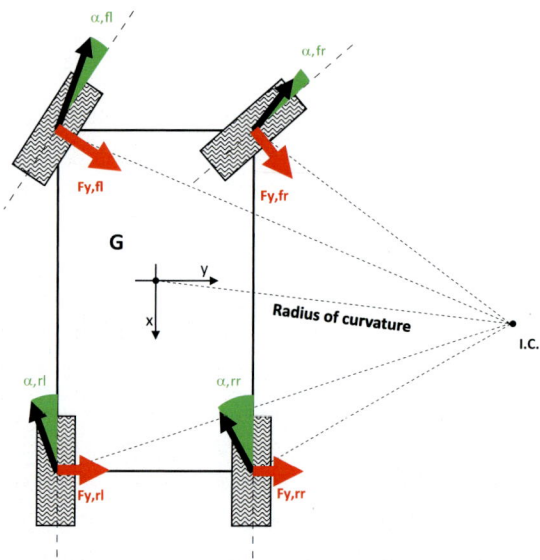

Figure 2 – Quadricycle in steady state condition. In red the forces, in green the slip angles. The force components do not indicate to the reference system of the car but to that of the tire.

These angles, identified with α in the previous figure, are the slip angles. They depend on stiffness of the tires and on the inertia of the vehicle to continue in the direction it had before steering was applied.

2.2. The slip angle

With reference to the front left wheel shown in figure 2, and enlarged in figure 3, the slip angle is the angle that the velocity vector of the wheel center forms with the longitudinal plane of the tire. As seen in the previous paragraph, the onset of a slip angle depends on the attempt by the vehicle, and therefore by the wheel center, to change direction. In particular, when a wheel turns, a transversal component of the wheel center velocity arises, which is due to the tire elasticity and directed towards the outside of the curve. Depending on how the tire is made, the creation of a slip angle induces the onset of a transversal force directed towards the inside of the curve which will balance, together with the force of the outer three tires, the centrifugal force acting on the vehicle.

Tire road interaction – Notes

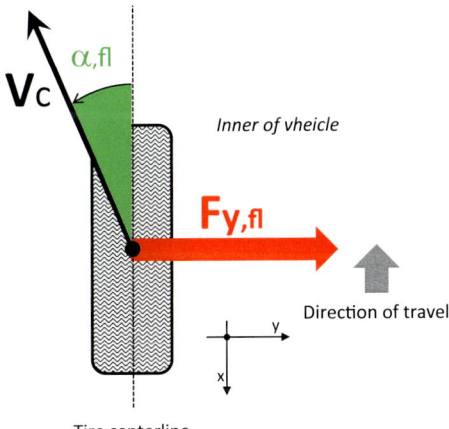

Figure 3 – Slip angle and relative transversal force of the front left tire. In the reference system both slip angle and relative transversal force are positive.

The generic transversal component *Fy* of the reaction force between tire and road, for small slip angles, is regulated by following linear law:

$$Fy = C_\alpha \alpha \quad (1.1)$$

where C_α represents the slip stiffness of the tire and therefore the slope in the origin of the curve of figure 4. It also turns out to be:

$$\tan \alpha = \frac{Vcy}{Vcx}$$

In the case of figure, α is positive because both *Vcx* and *Vcy* are negative.
The trend of the lateral force as a function of slip angle, in a steady state condition (with slip angle stabilized) is called *characteristic curve of the tire* and its shape is shown in the following figure:

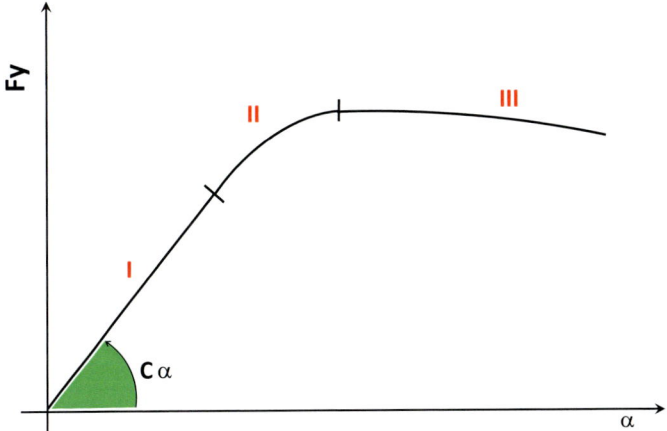

Figure 4 – Characteristic curve of the tire in the steady state condition.

Automotive suspension

The generic curve can be divided in: an initial linear section *(I)*, a central nonlinear but increasing section *(II)*, and a final either decreasing or constant section *(III)*.

Section *(I)* represents the initial linear behavior of the tire in which the force doubles by doubling the slip angle. If all the tires work in this section, we can say that the vehicle behavior is linear. To understand this last statement, which will be clearer in the next chapters, we must imagine the vehicle travelling through a curve with constant radius and increasing speed (the increase in speed is very gradual to allow for the vehicle to stabilize). In linear conditions, a constant increase in steering wheel angle corresponds to a constant increase in speed and therefore in centripetal acceleration.

Section *(II)* represents the non-linear behavior of the tire in which doubling the force triples or quadruples the slip angle.

Section *(III)* represents the limit zone where the tire reaches the maximum of the force with which it can react. This force, in correspondence with the subsequent slip angles, in some cases remains constant, in others it decreases. When we are in this section, even if we increase the slip angle, for example with steering, the reaction force does not increase. This means that the tire has reached its limit because its entire contact patch is in sliding condition and the vehicle cannot increase the lateral acceleration.

It is clear that, the more the C_α of tire increases, the more the vehicle is precise in change of direction. This is because with the same force, deriving from lateral acceleration, there is less slip angle and therefore a smaller deviation from the direction imposed by the tire. The characteristic curve described represents the trend of the force in steady state condition, but it does not take into account the transient state. Once a value of slip angle has been set, it is possible to trace the trend of the transient that brings the lateral force to the steady state value and get additional information on the tire response. The following graph shows the trend of the lateral force (relative to a certain slip angle α) as a function of the space traveled by the tire.

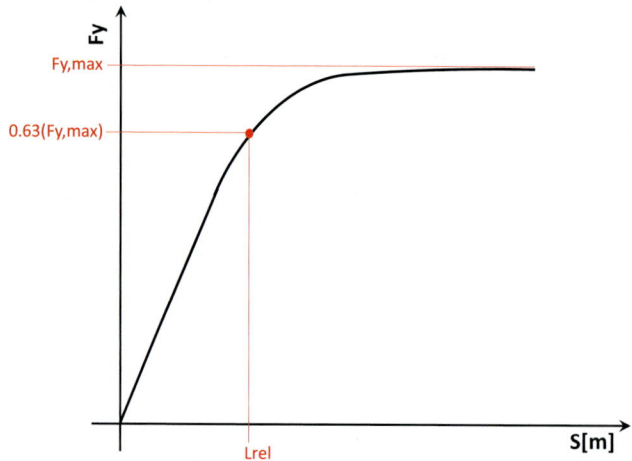

Figure 5 – Trend of lateral force as function of the space travelled by the tire with constant slip angle.

In the trend of the force, which asymptotically tends to the steady state value, the magnitude called relaxation length (*Lrel*), measured in meters, is highlighted. This is the space traveled by the tire to reach 63% of the maximum force, at constant slip angle. *Lrel* is inversely proportional to the response speed of the tire: the lower it is, the fastest the tire reaches the steady state force. Typically, low-profile tires (with low ratio height to width) have shorter relaxing lengths. On the vehicles which have different tire sizes between the axles, and which also certainly have different *Lrel*, it is advisable to combine these two values well, in order to have the desired lateral response of the vehicle. This topic will be discussed later.

2.3. The brush model

We have just discussed how the tire reacts under lateral load and how the steady state trend of the force is a function of the slip angle. This behavior can be explained by understanding the *brush model*. Only qualitative description of this model will be provided, not including the demonstration of some of the statements and referring to specialist texts for details. The brush model considers that the tire has an inextensible carcass, deformable only radially, and that the tread is made up of an infinite set of bristles. The bristles are considered with an infinitesimal length, zero mass, protruding radially from the carcass. The bristles, radially arranged on the entire width of the tread, are equipped with an extreme carcass side and an extreme road side and have two flexural stiffnesses *Kx* and *Ky*, in the longitudinal and in the transverse direction, respectively. In the following figure the schematization of the pure lateral interaction of a tire with wheel center velocity *Vc* and slip angle $\alpha 1$ is shown.

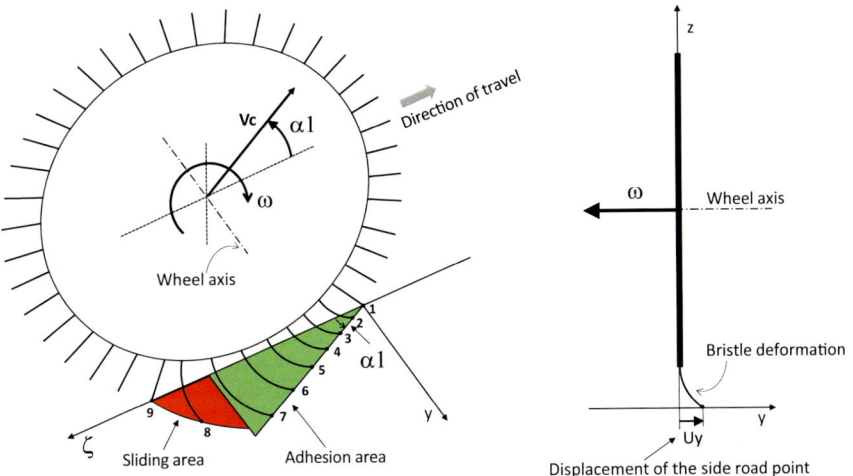

Figure 6 – Trend of transversal deformation of bristles of contact patch in the presence of a slip angle on a front tire left that curves to the right. The adhesion area in green and the sliding area in red. On the right (rear view) the deformation of a single bristle with the detail of the displacement *Uy* of the road side point.

Automotive suspension

On the left side of the figure the axonometric view (not perfectly drawn because adapted to the descriptive needs) of the brush model of a front left tire of a car that curves to the right is represented, with the trend of bristles deformation along the center line of the contact patch. The vector **Vc** belongs to the plane parallel to the road surface (π_ω) which contains the wheel axis. The lateral deformation of the generic bristle is represented on the right side, in an orthogonal projection, as looking at the wheel from the rear towards the front of the vehicle, with attention to the displacement of the point on the road side. The generic bristle, starting from the undeformed condition of the position no.1, assumes different deformed configurations (only transversely along y) as it passes through the contact patch that ends in position no.9. In particular, when the bristle is in positions between 1 and 7, the trend of the transversal deformation is linear along ζ and has slope equal to $\alpha 1$. In this region, also called adhesion area, the road side extreme of the bristle has zero lateral speed with respect to the road, and it is in static equilibrium since the return force due to the deformation is balanced by the static friction force. The elastic force acting on the bristle, which is dimensionally a tension, as it is expressed in N/mm², is:

$$\tau_y = K_y U_y$$

The elastic force during adhesion conditions is balanced by the static friction force of the road on the bristle, which depends from the vertical pressure value acting on the point in which the bristle is located. Therefore, it turns out to be:

$$\tau_y = K_y U_y = P(\zeta)\mu_{s,y}$$

as shown in the following figure:

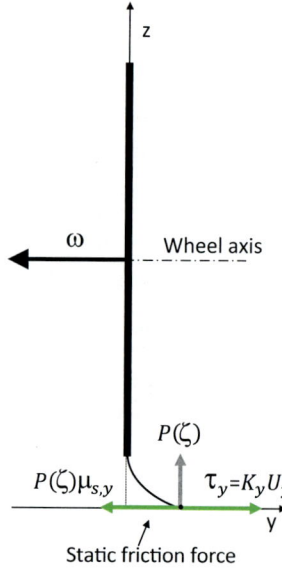

Figure 7 – Balance of the forces on the single bristle in adherence condition. The adherence condition persists until the friction coefficient reaches its maximum and then suddenly decreases to transform into dynamic friction.

Where $P(\zeta)$ is the value of the contact pressure as a function of ζ, dimension which represents, as shown in figure 6, the distance from the starting point (point no.1) to the generic point of the contact patch. The pressure trend, which in this modelling we suppose to be parabolic with the maximum corresponding to the center of contact patch, is as follows:

$$P(\zeta) = \frac{P_0}{a^2}(2a\zeta - \zeta^2)$$

In which P_0 depends on the vertical load as below

$$P_0 = \frac{3}{8ab} F_z$$

Where a and b are the half-length and the half-width of the contact patch, respectively. The vertical pressure trend, represented in the following figure, defines the trend of the maximum lateral static friction force (per unit area) that can be explicated on the generic bristle at distance ζ, which is equal to:

$$F_{max} = P(\zeta)\, \mu_{s,y}$$

Where $\mu_{s,y}$ is the maximum static friction between the bristle and the road.

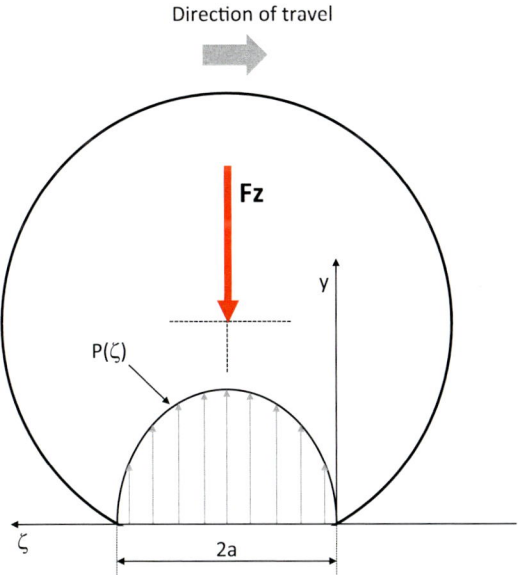

Figure 8 – Vertical pressure trend along the centerline of the contact patch in a lateral view.

When the bristle deformation reaches values such that the τ_y exceeds the maximum static friction force of that point of contact patch, the bristle begins to slide, the deformation decreases and the lateral force becomes:

$$\tau_y = K_y U_y = P(\zeta)\, \mu_{d,y}$$

Automotive suspension

This is lower than the previous one because the dynamic friction coefficient is smaller than the static one. Consequently, the deformation trend of the bristles, along the contact patch, undergoes a discontinuity while moving between the green and the red area. In the following figure the representation of the balance of the forces in case of sliding.

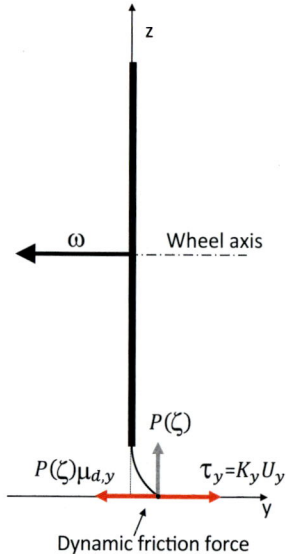

Figure 9 – Forces balance on the single bristle in sliding condition.

If the slip angle increases, the adhesion area decreases and the sliding area increases. This as shown in the following figure in which the increase in slope of deformations of the green area is evident.

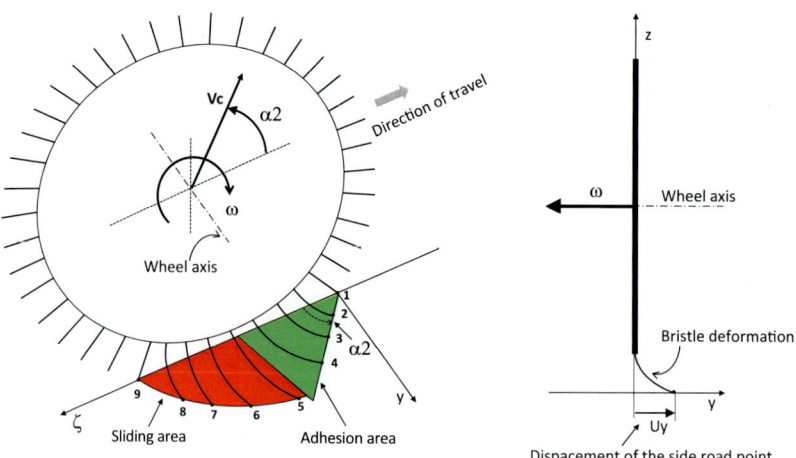

Figure 10 – Trend of the transversal deformation of the bristles of the contact patch in the presence of a slip angle greater than the previous one. On the right, the deformation of the single bristle.

The contact area between tire and road, in presence of a slip angle, is called pseudo-sliding area because it is constituted by points where the deformed bristles don't slide and points where the deformed bristles slide. The first points constitute the adhesion area and the second the sliding area. When the slip angle increases, the sliding area increases, and the adhesion area decreases.

To map all the lateral stresses inside the contact patch we only need to multiply the deformations U_y of the figure 6 and 10 by the transverse stiffness K_y. Below the aforementioned trend in function of ζ. In particular, it can be seen that the transverse stresses grow linearly and reach their maximum in correspondence of the length l_a, called *adhesion length*, when the transverse stress is equal to the maximum static friction force explicable by the bristle. For $\zeta = l_a$, the lateral stress acting on the bristle undergoes a discontinuity because it switches from a condition of static equilibrium P_s to a condition of dynamic equilibrium P_d. For $\zeta > l_a$, the bristle is in a sliding situation and the stress acting on it is a dynamic friction force that depends on the vertical pressure trend $P(\zeta)$. This because it is the product between the latter and the corresponding friction coefficient. The stresses graph across the length of the contact patch is described by the green line until the point P_s, and by the curved line in red from the point P_d onwards.

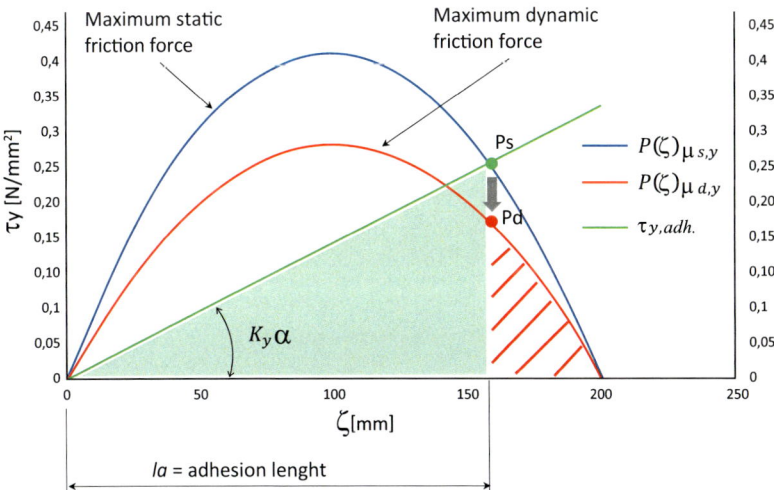

Figure 11 – Graphic representation of the transversal stresses trend. The green line represents the tensions in the adhesion area while the curved line, from the point P_d onwards, represents the trend of the tensions in the sliding area.

The integral of the transverse stresses extended to the length of the contact patch is represented by the sum of the area in solid green and of the hatched area in red. By multiplying this sum by the contact patch width, we obtain a good approximation of the lateral force relative to the corresponding slip angle. Increasing the slip angle, the slope of the linear part of deformations increases (green line), and consequently the distribution of the areas to be added to obtain the lateral force varies. In the following figure it is shown the graph of the stresses corresponding to a greater slip angle which causes the decrease of the *adhesion length*. The sum of the areas is greater than in the case of the previous figure and therefore also the lateral force is greater, but it is foreseeable that the increase in force as a function of the slip angle is in a decreasing phase. In fact, in figure 13 it can be seen that, once the slip angle which brings the green curve to be tangent to the blue curve has been reached, the lateral force is defined only by the dashed area which remains constant in increasing of slip angle. The tangent to blue curve in the origin will be called *the sliding limit*. This describes, with good approximation, the trend of the lateral force as a function of slip angle in section following the linear one. However, this does not take into account the decreasing section of the graph.

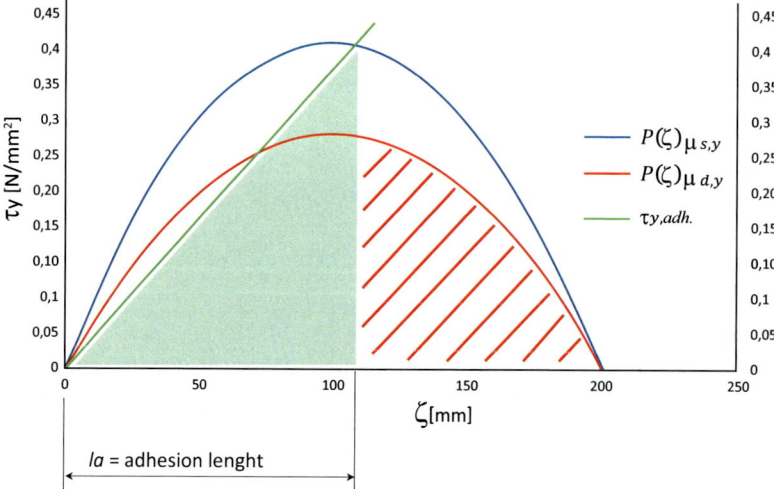

Figure 12 – Transverse stress graph corresponding to a greater increase in the slip angle than in the previous case.

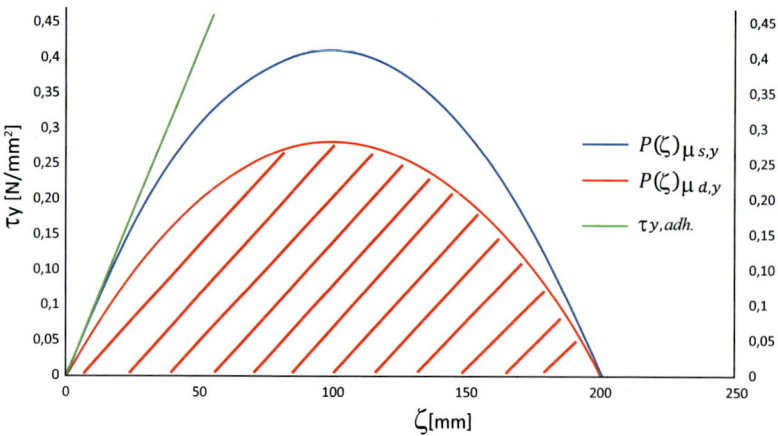

Figure 13 – Transverse stress graph relating to the crisis slip angle that cancels the adhesion area. The red dashed area represents the transverse slip stresses.

On the other hand, when α is small, the trend of the force is linear. In this condition, as shown in the following figure, the green triangle area is predominant with respect to the red dashed one, which becomes negligible. When the slip angle is small, the base of the triangle can easily be approximated to $2a$, while the height h is directly proportional to the slip angle according to the following relationship:

$$h = 2aK_y\alpha$$

the area of the triangle is:

$$A = \frac{1}{2} 4a^2 K_y \alpha$$

and the transverse force is equal to the product between the area and the depth $2b$ of the contact patch, which for small angles is proportional to α as below

$$F_y = 2a^2 2bK_y\alpha$$

For this reason, it can be asserted that, for small angles, the tire characteristic is linear and its stiffness in the origin is equal to:

$$C\alpha = \frac{F_y}{\alpha} = 2a^2 2bK_y$$

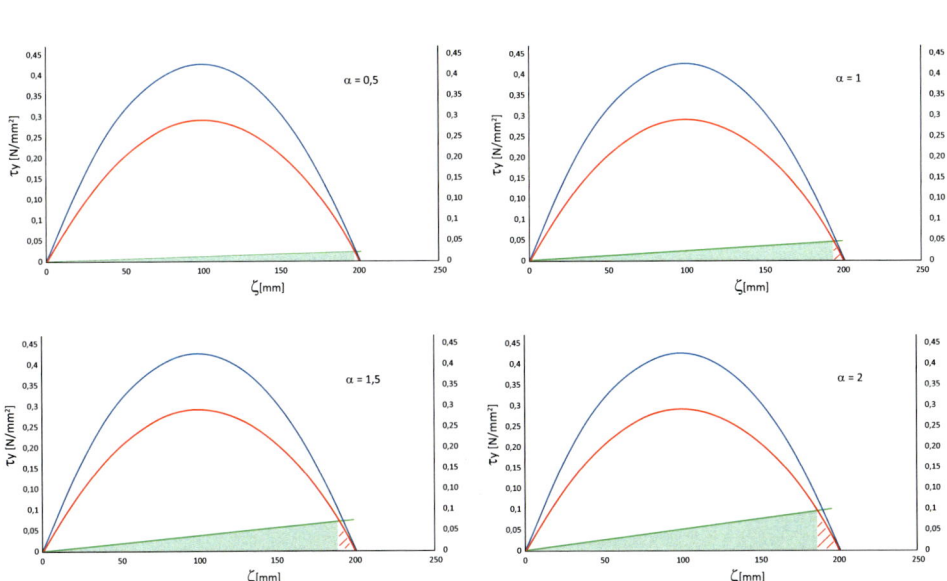

Figure 14 – Distribution of areas corresponding to the transverse forces relative to small slip angles ranging from 0.5 and 2 degrees.

2.4. Camber effects on the tire characteristic

An effective and immediate way to modify the trend of tire characteristic of figure 4 is to vary the *camber* angle.

The *camber* is the angle between the longitudinal plane of the tire and the vertical plane. It is negative when the distance between the wheels measured on top is less than that measured on the road, as shown in the following figure:

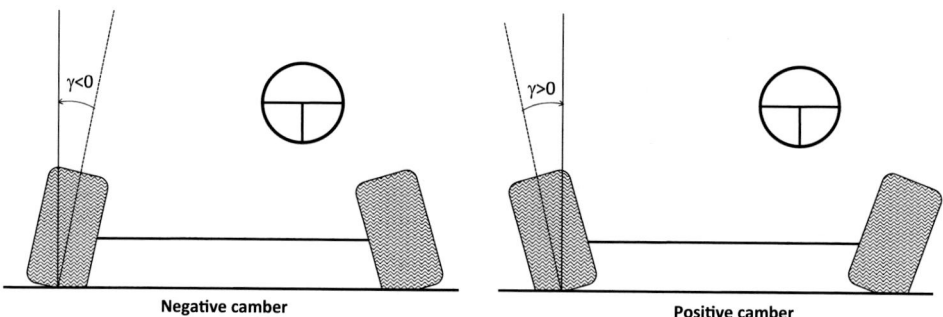

Figure 15 – Front view of the *camber* angle negative on the left and positive on the right[1].

The presence of a *camber* angle different from 0 produces the transversal displacement of the side carcass points and the consequent deformation of the bristles, as shown below:

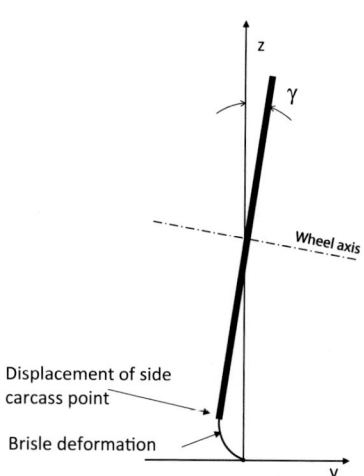

Figure 16 – Deformation of the single bristle due to the *camber* < 0 in a front view.

In the figure it can be seen that, during the application of a vertical load to a wheel with negative *camber*, the extreme side road of the bristle is fixed while remaining in the

---

1. In all schematics of the book there are only left-hand drive car.

original position (stuck to the ground), while the one on the side carcass deforms transversely towards the outside of vehicle. The elastic configuration assumed by the bristle generates a force on the tire directed towards the inside of the vehicle (*camber<0*). By superimposing the elastic deformation due to a *camber<0* to the deformation due to slip angle, the situation described in the following figure is obtained. On the left side of the figure, the parabolic trend of the deformation of the carcass side points due to the *camber* is shown in purple (as a function of ζ), while on the right side the overlap of the negative *camber* deformation and the slip deformation on the single bristle is shown. From the observation of the deformations and the consequent forces, it is very clear that the negative *camber* produces a transverse force on the tire directed towards the inside of the vehicle which is added to that due to the slip angle. The most significant consequences of the negative *camber* are:

1. With the same slip angle, the tire with *camber<0* reacts with a greater transverse force.
2. The generic bristle of the contact patch undergoes greater deformation and first reaches the condition of discontinuity between adhesion and sliding. Consequently, with the same slip angle, lenght *la* decreases. The decrease of the magnitude *la*, for each slip angle, is responsible for the variation of the trend of tire characteristic. In particular, the *sliding limit*, namely the slip angle corresponding to the maximum lateral force, decreases.
3. With the same slip angle, the wear of the tread increases due to the increase in the sliding area. This represents one of the limits to the increase of *camber*.

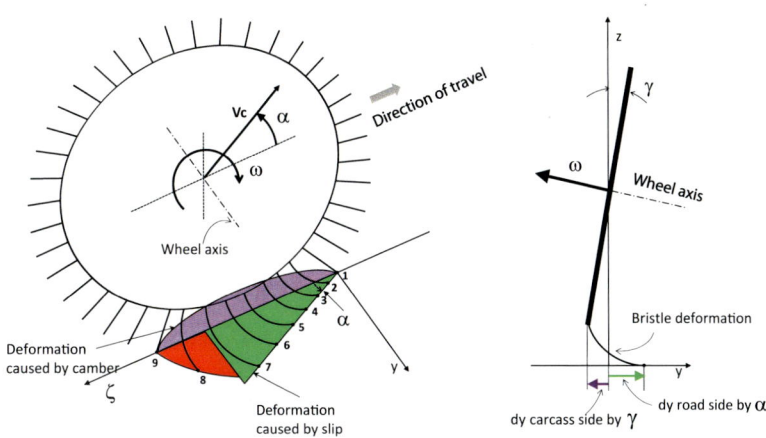

Figure17 – Overlap of the *Uy* deformations due to *camber<0* and slip angle on a front left tire which curves to the right.

From what has been said it follows that the qualitative trend of the tire characteristic as a function of *camber* is as shown in the following figure, where the characteristic of a tire mounted on a suspension which gains *camber* in bump travel is also shown. In the latter case the variation (dashed curve in purple) does not manifest itself immediately, but only when the wheel begins to gain *camber* due to the suspension bump travel.

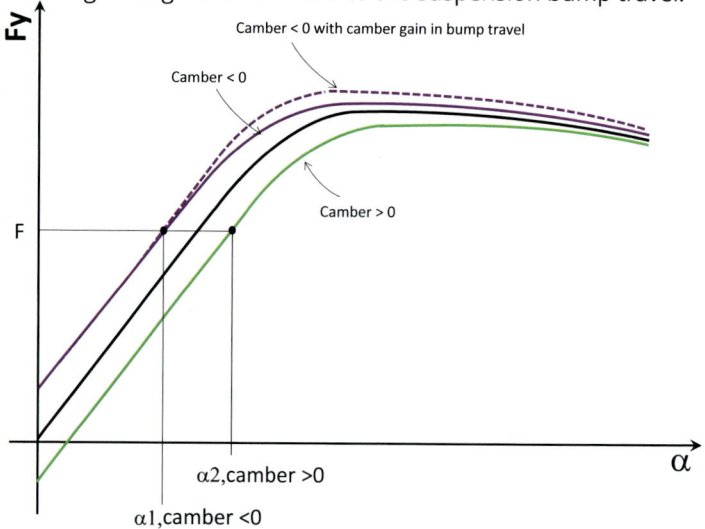

Figure 18 – *Camber* effect on the tire characteristic. The *camber* gain due to suspension kinematics further moves the curve towards left.

In the presence of *camber*, the 1.1 becomes:

$$Fy = C_\alpha \alpha + C_\gamma \gamma$$

Due to the sign convention used for the *camber*, C_γ takes negative values.
From the previous figure it can be seen that, with the same lateral force, a tire with greater *camber* (negative) reacts with a lower slip angle. This leads to a greater directional precision with respect to a tire with *camber* 0. Another way to interpret the characteristic with negative *camber* is the increase in lateral force with the same slip angle imposed by the steering, which increases the yaw moment around the vertical axis, giving the vehicle greater readiness when cornering. As previously anticipated, the limit of this phenomenon lies in the fact that the tire characteristic, in the presence of the *camber*, not only undergoes a translation but, due to the decrease in the surface of the contact patch and the reduction of *la*, also a shape change. In particular, when the *camber* is negative, the linear section *(l)* reduces and the *sliding limit* (slip angle relative to the maximum force) decreases. Furthermore, by increasing the *camber* angle, in addition to having uneven wear of the tread, the resistance to advancement increases.
The influence of the *camber* on tire behavior can be motivated also by more practical considerations. In fact, in case of negative *camber* the lateral force, emerging from the direction change, increases the thrust along the longitudinal plane of the tire. This, in turn, modifies the lateral characteristic making the tire more reactive. In the following figure it is shown the breakdown of the forces highlight that the component *F2* increases the transverse stiffness of the carcass which is laterally stressed from the component *F1* which is less than the lateral load.

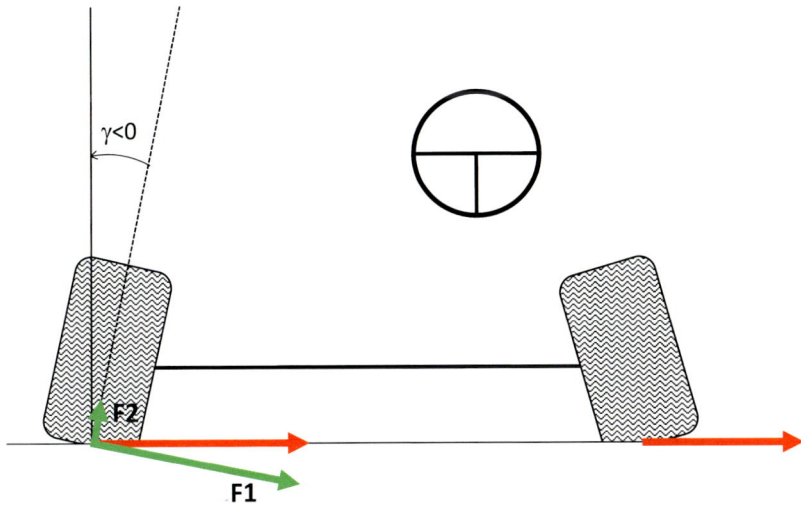

Figure 19 – Breakdown of the lateral force acting on the tire in presence of negative *camber*.

In race cars, where it is more important to have the maximum of transverse thrust to reach high lateral acceleration with little slip angles, it is advisable to increase the *camber* as much as possible having as limits both the resistance to advancement and the braking performance which decreases with the contraction of contact patch. In road cars the focus is on the uniform wear of the tires.

What postulated above is true because the wheel outside the curve has a predominant role on the axle behavior, since the negative *camber* on the inner wheel induces an opposite behavior because it lowers the maximum lateral thrust of the latter.

2.5. Other effects on the tire characteristic

The other quantities that influence the tire characteristic are:
- The coefficient of adherence;
- The inflation pressure;
- The vertical load.

2.5.1. The coefficient of adherence

The coefficient of adherence affects the characteristic of the tire because it arises from the behavior of the forces of interaction between the tire and the road. The coefficient of adherence depends on the coefficients of friction (static and dynamic) of the rubber with the road, on the pressure inside the contact patch, on the coexistence of loads in both directions (longitudinal and transverse) and on the stiffness of the tread compound. In order to describe the influence of some of the aforementioned parameters, we note that a rubber with soft compound penetrates more easily into the micro cavities of the surface of the road, increasing the number of elements participating to the lateral resistance; when, on the other hand, the friction coefficient decreases, the parabolas of *figure 11* are lowered, causing a decrease in the *sliding limit* and the maximum lateral force. Below is the trend of the variability of the characteristic as the coefficient of adherence μy varies.

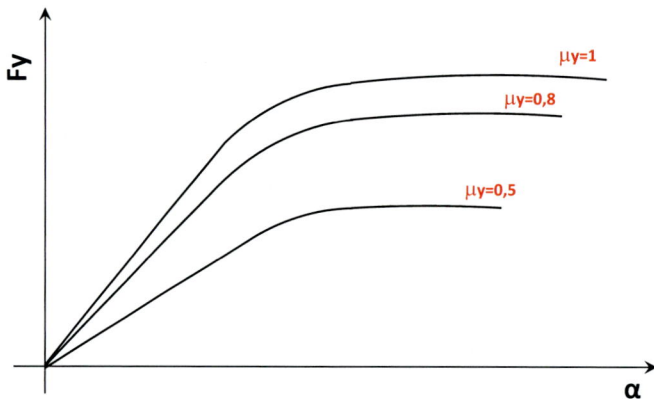

Figure 20 – Tire characteristic trend as the coefficient of adherence.

The adherence coefficient is the rate between the lateral force and the vertical force acting on the tire and represents the percentage of the vertical force which determines the lateral force. It is maximum when it is corresponding to the peak of the lateral force.

$$\mu y = \frac{Fy}{Fz}$$

The value of μy depends also on the load conditions at ground because when the tire is loaded also in longitudinal direction the maximum value of μy decreases.
Let μx be the longitudinal adherence coefficient for which the following holds:

$$\mu x = \frac{Fx}{Fz}$$

where the *Fx* is the driving or braking force.
We shall now introduce the concept of adherence capital which describes that the maximum load, which the tire can support in one direction, decreases as the applied load in the orthogonal direction increases.
In this regard, we can define the adherence ellipse of the tire through which it is possible to trace the maximum value of the adherence coefficient in conditions of mixed use of the two directions.

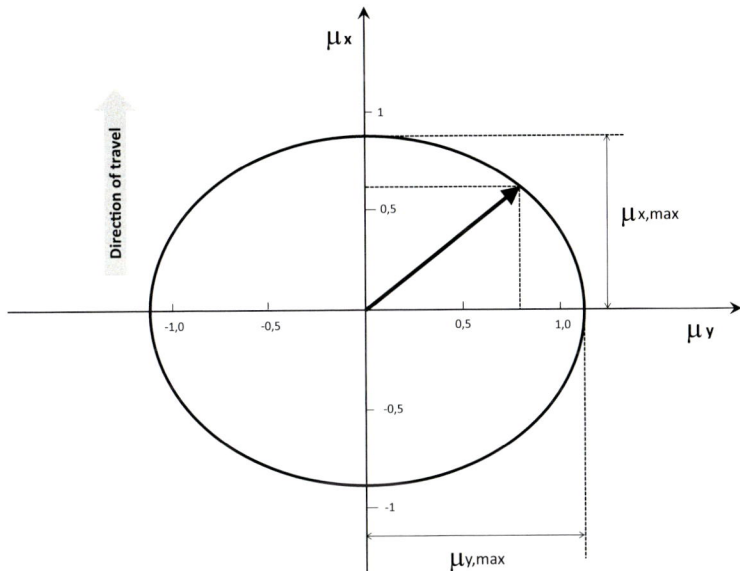

Figure 21 – Adherence ellipse of a road tire.

As shown in the figure, if the tire is subjected to a traction force that engages a coefficient of adherence equal to *0.6*, the maximum adherence coefficient in the transverse direction is *0.8* and therefore the maximum lateral force before sliding is *0.8Fz*. The following figure shows the elliptical trend of the road tire interaction force.

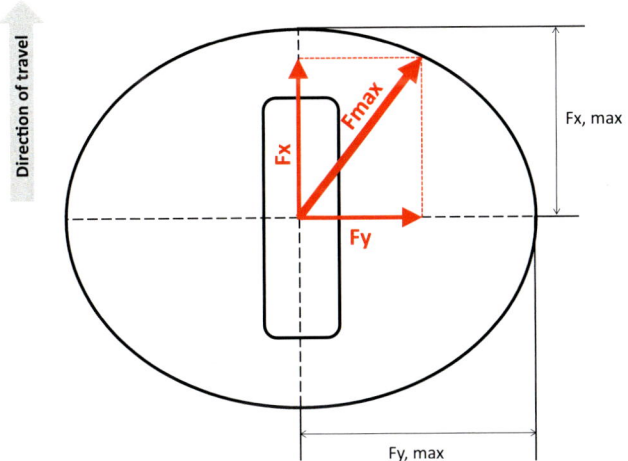

Figure 22 – Trend of maximum lateral force as function of longitudinal force used.

The maximum lateral force, which represents the maximum of the tire characteristic curve, depends on the longitudinal force used for traction or braking. From this it can be seen that the driving axle tires undergo a lowering of the lateral characteristic curve and consequently, with the same transverse force, they generate greater slip angle. In the following figure an example of how the lateral characteristic varies in the presence of traction on the axle with the evidence of the lowest *sliding limit ($\alpha max,2$)* is shown. The extreme case of this behavior occurs when, in correspondence with an energetic acceleration, which uses the maximum available longitudinal adherence, the vehicle is unable to face a curve because the transverse reserves of adherence are canceled.

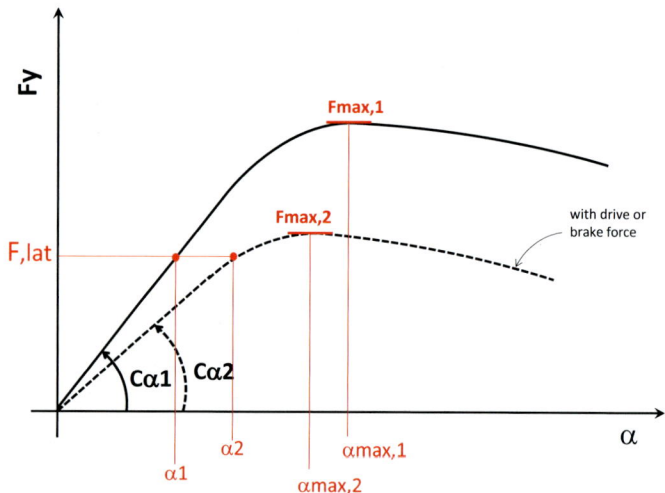

Figure 23 – Lateral characteristic of a tire with and without traction. With the same transverse force, the slip angle increases in the presence of a longitudinal load. In this case the *sliding limit* is lowered (slip angle corresponding to *Fmax*).

The same observations apply in presence of longitudinal braking loads.
From what has been said, it can be stated that, to obtain the maximum performance in traction or braking, the wheels must have *camber* and *toe* equal to zero. Otherwise, transverse forces would be triggered which, in turn, would lower the adherence capital in the longitudinal direction.
Another aspect that should not be overlooked is the variability of the adherence coefficients as the vertical load acting on the tire varies. This phenomenon, not very evident for low loads, consists in the decrease of the adherence coefficient as the vertical load increases. The reason for this last dependence lies in the fact that the consequent increase of the pressures inside the contact patch produces an alteration of their uniformity (previously they were supposed to be parabolic only for simplification purposes) with consequent degradation of the interaction between the tire and road. We can, therefore, state that the adherence coefficients increase when the pressures inside the contact patch are lowered. From this it follows that wider tires, which have larger contact patch, produce, with the same vertical load, an improvement in both longitudinal and lateral adherence. This last statement is not always true on very low adherence surfaces (snow or mud where the tire must "claw" at the road surface) but, since these extreme conditions will not be addressed, the more detailed explanation is left to the specialized texts.

Special attention should be paid to the fact that the lowering of μx and μy, as function of vertical load, does not imply lower longitudinal and transverse loads but a lower ratio between the latter and the vertical load.

2.5.2. The temperature and the inflation pressure

The inflation pressure affects the characteristic curve because it is responsible for the size and shape of the contact patch and therefore for the vertical pressure distribution within it. By increasing the inflation pressure, the contact patch area decreases and consequently the interaction capacity between the tire and the road varies. Unfortunately, it is not easy to formulate a theory capable of identifying the correct value to be applied. This is because the pressure increases as the operating temperature of the tire increases, which, in turn, depends both on the use and on the external temperature. Furthermore, since the transverse stiffness of the tire is proportional to the pressure, we can say that a tire with low pressure tends to warm up more due to the greater deformations induced.

In tires used in competitive events, the evaluation of the correct inflation pressure is made according to the measurements of the temperatures of the inner, central and outer tread band. The measured values provide an indication of the correctness of the pressure and *camber* used. For optimal operation it is preferable to have the temperature of the three bands within a range provided by the manufacturer, with an increasing trend going from the outside towards the inside of the tire. The central band must remain at the center of the range, while a positive ΔT must be added to the inside band and a negative ΔT must be added to the outside band, as shown in the following figure. The ΔT ranges from 5 to 10 degrees depending on whether the tire is road or slick.

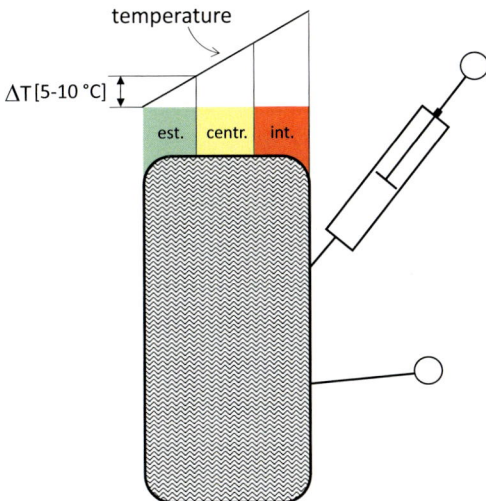

Figure 24 – Trend of tread temperature.

The rules to follow to set the correct inflation pressure and the correct static and dynamics (gain) *camber* are as follows:
- A higher than specification temperature in the central band and lower in the lateral bands indicates that the inflation pressure is too high.
- A higher than specification temperature in the lateral bands and lower in the central bands indicates that the inflation pressure is too low. This also implies a greater warm up of the tires due to the transverse stiffness lowering.
- A higher than specification temperature in the internal band indicates that the *camber* (negative) is too high.
- A higher than specification temperature in the external band indicates that the *camber* (negative) is too low.

The following table summarizes what has been said. The temperature greater or lower than target value is indicated with an up or down arrow. Remember that too high *camber* means "too negative".

External temperature	Central temperature	Internal temperature	Camber [OK/NOK]	Pressure [OK/NOK]
M-5°	M	M+5°	OK	OK
↓	↑	↓		NOK (High)
↑	↓	↑		NOK (Low)
↓	M	↑	NOK (High)	
↑	M	↓	NOK (Low)	

M = Center of the range supplied by the manufacturer.
Figure 25 – Guideline for *camber* and inflation pressure set-up.

During these adjustments, it is important to keep in mind that, when the inflation pressure is lowered, the adherence of the tire tends to increase but at the same time the lateral stiffness, that opposes the loss of *camber*, tends to be lowered as well. Consequently, a lowering of the pressure may require an increase in the static *camber*, in order to stem the possible increase in the external temperature caused by the lateral deformation of the tire. Amongst the causes of the rise in tire temperature there are the *toe* and the vertical load. In particular, the opened wheels (*toe-out*) increase the temperature of the inner side of the tread while the closed wheels (*toe-in*) increase the temperature of the outer side. The dependence on the load mainly concerns the values of the transfers when cornering, braking or accelerating and the speed with which the vertical force increases which, in turn, depends on the stiffness of the suspension on the ground, a characteristic that will be clarified in the next chapter.

2.5.3. The vertical load

The vertical load is a frequently variable quantity due to the continuous load transfers that occur during use of the vehicle. It has a fundamental role in the operation of the tire because it influences the specific pressure inside the contact patch that, in turn, determines the maximum lateral and longitudinal forces.

It is quite intuitive that, by increasing the vertical load, the whole body of the tire, being more compressed, opposes greater resistance to the attempt of deforming it laterally. In fact, experimentally it can be seen that the increase of vertical load generates an increase in C_α, an upward shift of the entire tire characteristic and the increase of the *sliding limit*. The aforementioned increments are not progressive but, as shown in the following figure, decrease as the load increases and the upward shift of tire characteristic is not proportional to the *ΔFz*. This peculiarity is of fundamental importance for understanding the behavior of the entire axle under lateral load.

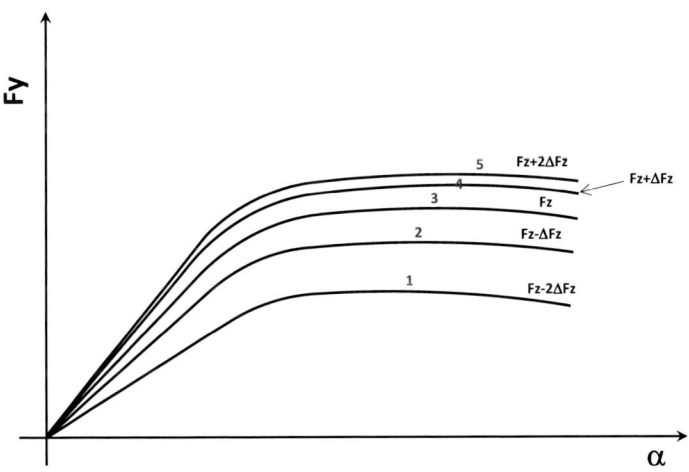

Figure 26 – Dependence of lateral characteristic on vertical load.

2.6. The load transfer and the slip angle

In a steady state condition, the average value of the slip angle of the two wheels of the same axle depends on the load transfer due to the centrifugal transverse force. This on the assumption that the vehicle, in its cornering motion, has large radii of curvature and low steer angles. In this condition we can say that, during cornering, the slip angles of the two wheels of the same axle are almost equal. We shall demonstrate the latter statement by referring to the axle of the following figure, in which the load transfer forces due to the lateral centrifugal force *Y* are represented.

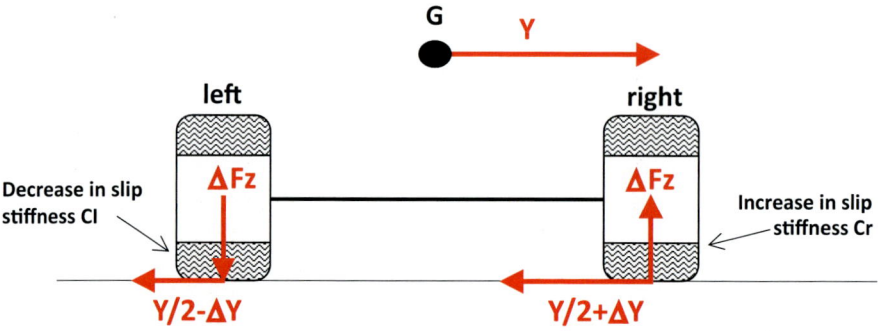

Figure 27 – Representation of vertical and lateral load transfer.

In order to make a qualitative estimate of the $\varDelta Y$, we suppose that the lateral stiffness of the tire, namely the load required to move the wheel center along y by a unitary quantity, is proportional to the slip stiffness. This hypothesis is not always verified because the lateral stiffness depends on the carcass structure while the slip stiffness depends also on the coefficient of adherence. Having said this, we observe that, since the system is hyperstatic along the transverse direction, the distribution of forces on the two wheels depends on the distribution of the lateral stiffnesses which are different due to the load transfer. In fact, the stiffness Cr increases due to the vertical compression of the tire owing to the load increase, while the Cl decreases due to the decrease in load. Where the slip stiffness increases, the transverse force also increases because, with the same transverse deformation, the stiffer wheel reacts with greater force. By applying the principle of virtual works, we obtain that the distribution of the forces follows the distribution of the stiffnesses and therefore:

$$\frac{Y}{2} - \varDelta Y = Y \frac{Cl}{Cl + Cr}$$

$$\frac{Y}{2} + \varDelta Y = Y \frac{Cr}{Cl + Cr}$$

By subtracting the first equation from the second we get:

$$\varDelta Y = \frac{Y}{2} \frac{(Cr - Cl)}{(Cl + Cr)}$$

The slip angles, in the hypothesis of linear behavior of the tires, will be:

$$\alpha_{,}l = \frac{Y/2 - \varDelta Y}{Cl} = Y \frac{Cl}{Cl + Cr} \frac{1}{Cl} = \frac{Y}{Cl + Cr}$$

$$\alpha, r = \frac{Y/2 + \Delta Y}{Cr} = Y \frac{Cr}{Cl + Cr} \frac{1}{Cr} = \frac{Y}{Cl + Cr}$$

and then

$$\alpha, l = \alpha, r$$

This equation was verified (with a non-rigorous demonstration) by involving forces and stiffnesses to provide a qualitative assessment of the phenomenon of lateral force transfer; the most rigorous demonstration will be proposed at the beginning of chapter 4. By exploiting this equation, through the graph of figure 26, it is possible to build the characteristic of the entire axle by adding, for each slip angle (common to the two wheels), the forces relative to the two curves which correspond to the same load transfer. Below is the construction described related to the different load transfers.

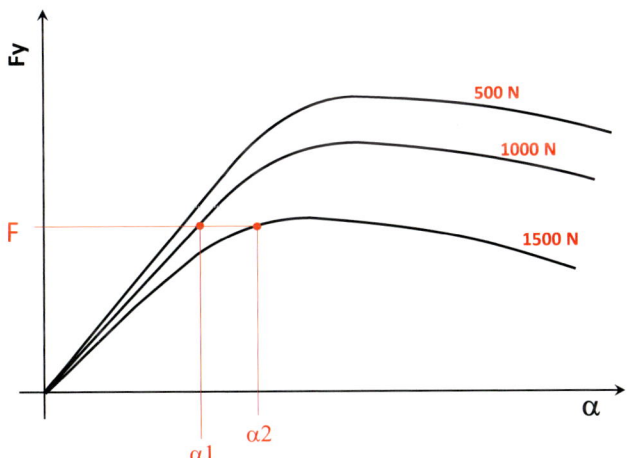

Figure 28 – Characteristics of the axle as the load transfer varies.

The graph shows that the more load transfer increases, the more the axle characteristic is lowered. In fact, if F is the centrifugal force distributed on the axle, the *1500N* load transfer curve will correspond to an angle $\alpha 2$ greater than angle $\alpha 1$ corresponding to the *1000N* curve. We can conclude that, in stationary conditions, with the same transverse force, which depends only on the mass and the mass distribution, the slip angle of the axle increases with increasing load transfer. If, for example, we want a greater slip angle on the rear axle to decrease the radius of curvature, we must increase the load transfer. It is important to reiterate that the *Fy* does not depend on the load transfer. Later we will see what the techniques are to increase the load transfer and what is the correct meaning of variation of radius of curvature.

2.7. Summary

In conclusion, based on what has been said in the previous paragraphs, we can summarize everything in two points:
- The direction of velocity of center of wheel is never parallel to the longitudinal plane of the tire but it is always directed towards the outside of the curve with an angle called slip angle.
- The emerging transverse force is always directed towards the inside of the cornering. In steady state condition, it has a linear trend for small slip angles, reaches a maximum and then decreases. The trend of the transverse force as function of slip angle depends on:
    - The available adherence. By lowering the adherence coefficient, the values of the force that can be exerted by the tire for each slip angle are lowered. The characteristic curve (the lateral maximum force) lowers in case of braking or driving.
    - The vertical load on the tire. The characteristic curve increases with the load. As the vertical load increases, the slip angle corresponding to a certain lateral force decreases and the slip angle corresponding to the crisis (sliding limit) increases. On the contrary, with the same transverse force, the slip angle of the axle increases as the load transfer increases.
    - The *camber*. As the *camber* increases, the slip angle corresponding to a certain lateral force decreases. The additional lateral thrust, determined by the *camber*, lowers the braking and traction adherence.
    - The temperature and the inflation pressure. The inflation pressure together with the static *camber* contribute to the heating of the tire. For an ideal tire operation, the tread temperature must be decreasing towards the outside of the vehicle and the average value must be the one suggested by the manufacturer.

3. THE MOTIONS OF THE VEHICLE

3.1. General considerations

To analyze the dynamic behavior of a vehicle, it is necessary to make preliminary considerations on the possibilities of relative movement between the various subsystems that constitute it and, in particular, between the vehicle body and the wheels. These motions are determined by the different drive conditions and are the natural consequence of applied forces. The different loading forces commonly considered in the dynamic study of a vehicle are inertia forces, aerodynamic forces, longitudinal and transverse forces of tire-road interaction.

Before undergoing a detailed dynamic study, it is required to make a description of moving masses in order to identify both the exchanged forces and the position assumed by tires in the operating conditions of the car (*characteristic angles*).

3.2. Sprung and unsprung masses

The total mass of the vehicle can be divided, for the purpose of studying its dynamic behaviour, in sprung masses and unsprung masses connected to each other through a suspensions system. We will describe below the main components of the sprung masses and those of the unsprung masses.

Within this distinction, the engine, gearbox and passengers are considered integral with the vehicle body and contribute, together with the latter, to form the so-called sprung mass of the vehicle. To study the global dynamics of the system, the sprung mass of the vehicle is considered as a perfectly rigid structure. This hypothesis is certainly debatable because the bodies and frames of the vehicle are not rigid. However, introducing the flexibility of the sprung mass would make the dynamic analysis too complicated and would slightly modify the final results.

The elements of the unsprung mass are constituted by the four wheels, a part of the suspension system and everything that is anchored with the upright such as discs, calipers, etc.. The suspension system is schematized by adding 1/3 of the masses of the elements connected to the vehicle body to the unsprung mass; for example, the lower control arm will have the 2/3 of its mass added to the vehicle body and 1/3 added to the upright (unsprung mass). The most effective schematization foresees an unsprung mass for each wheel, as shown in the following figure in which only the left side of the vehicle is represented. Each of the four unsprung masses is constrained to move, relative to sprung mass, according to the motion law imposed by the suspension kinematics. The reaction forces to these movements come from the stiffness $Ksusp$, derived from the springs, and from the damping C of shock absorbers. The unsprung masses are connected to ground through the tire stiffness Kt which generally have one additional order of magnitude than those of the springs. The described schematization, which concerns the suspension systems with independent wheels, must be revised if there is an axle with dependent wheels. In this case the axle will have only an unsprung mass, equal to the sum of the left and right ones, connected with two springs and two dampers to the sprung mass. It's clear that in the presence of an axle with dependent wheels, the relative movement of the unsprung mass will have a much more evident effect than in the case of independent wheels, both for the connection of the two wheels and for the greater mass involved. In general, the unsprung mass value is an order of magnitude smaller than the sprung mass.

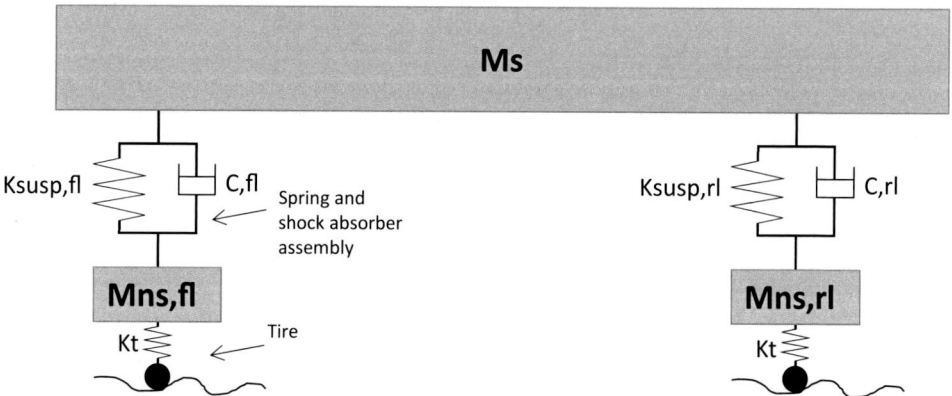

Figure 1 – Side view of vehicle schematic with sprung mass Ms, unsprung masses Mns, ground suspensions stiffness $Ksusp$ and tires stiffness Kt.

3.3. The pitch and rebound motions

The pitch and rebound motions are the most important relative movements in the study of vehicle *comfort*.
The pitch is the rotation motion of the car around a transverse axis, called pitch axis, which generally does not pass through the center of gravity.

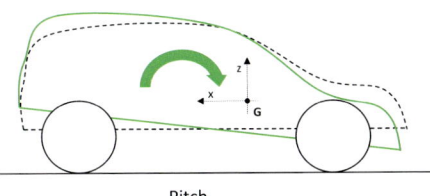

Pitch

Figure 2 – Pitch motion of the sprung mass.

The oscillation frequency of this motion, in addition to depending on moment of inertia with respect y, is a function of the suspension stiffness at ground and of the distance of the latter from the pitch axis.
The rebound is the translation motion of the car in the positive vertical direction (extended springs) and takes the name of bump when the direction of the motion is negative (compressed springs). Generally, the term "rebound" is used to indicate the vertical motion, regardless of the direction. The free oscillation frequency, in addition to depending on the value of sprung mass, is only a function of the suspension stiffness at ground and not of their relative longitudinal position.

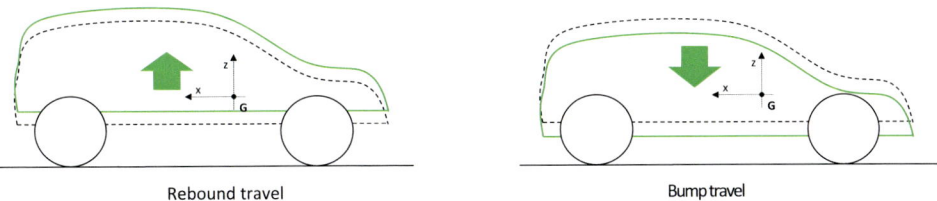

Rebound travel   Bump travel

Figure 3 – In the rebound motion (on the left) the suspensions extend while in the bump motion (on the right) the suspensions compress.

From the studies carried out it has been seen that, to have an optimal condition of *comfort*, the free frequency of the two motions must be contained in an interval between 1 and 1.5 Hz. A practical explanation of this statement lies in the fact that the frequencies included in that range are typical of the human step and therefore of our habits. The aforementioned frequencies can increase by one or two points in the case of sports cars where *comfort* is less important.
Once the designer identified the optimal frequencies of rebound and pitch, they must know how to manage the position and the stiffness of the springs to obtain these values.

3.4. The bump stiffness

The bump stiffness, evaluated on the single wheel, is the value of the force necessary to compress the wheel one millimeter upwards, keeping the vehicle body fixed. It is measured in N/mm and together with the sprung mass helps to define the bump frequency f (called also rebound frequency):

$$f\,[Hz] = \frac{1}{2\pi}\sqrt{\frac{K_{susp}}{M_{sem}}}$$

Where K_{susp}[N/mm] and M_{sem}[kg] are respectively the bump stiffness at ground and the sprung mass of the semicorner (the half of the sprung mass of the axle) described in the following figure. By imposing the frequency value it is possible to calculate the suspension bump stiffness K_{susp}, which represents the start point for calculating the characteristics of the spring.

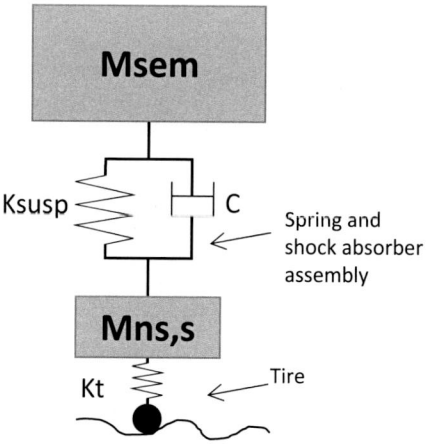

Figure 4 – Schematization of the generic semicorner with sprung masses (M_{sem} is equal to half the sprung mass of the axle), unsprung masses ($M_{ns,s}$), suspension stiffness K_{susp} [N/mm] and tire stiffness K_t [N/mm].

The ground stiffness of the suspension K_{susp} is directly proportional to the stiffness of the spring K_{spring}. The relationship between the suspension stiffness at ground and spring stiffness depends on the *Lambda* ratio between the displacement S_s of the spring (see figure below) and the displacement S_g of contact point at ground (approximately at the center of the tire):

$$K_{susp}\,[N/mm] = K_{spring} Lambda^2$$

where *Lambda* is:

$$Lambda = \frac{S_s}{S_g}$$

The *Lamba* ratio, which is generally less than 1, links also the ground load F_g with the spring preload *Fs* according to the following relationship:

$$F_g [N] = F_s Lambda$$

and then

$$F_s [N] = F_g \frac{1}{Lambda}$$

Where *Fg* is the ground force which equals the weight force of the sprung mass of the semicorner while *Fs* is the preload which the spring must have to statically balance this force.

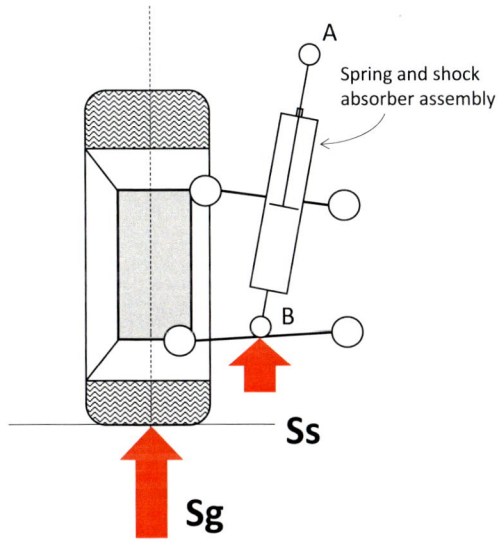

Figure 5 – Lambda ratio = Ss/Sg.

These relationships are obtained from the equality between external work, carried out by the ground force, and internal work carried out by the spring, as follows:

$$Le = F_g S_g = Li = F_s S_s$$

from which it results:

$$F_g S_g = F_s S_s$$

and than

$$F_g = F_s \frac{S_s}{S_g} = F_s Lambda$$

By replacing the forces with the respective products between stiffness and displacement we obtain:

$$F_g = K_{susp} S_g$$

which inserted in previous formulations

$$K_{susp} S_g = K_{spring} S_s \frac{S_s}{S_g}$$

they lead to the used relationship

$$K_{susp} = K_{spring} Lambda^2$$

3.5. The sizing of the springs

Sizing a suspension spring means defining its stiffness and free length so that, installed on the vehicle, it gives it the desired height from the ground. In the spring sizing we refer to the semicorner. The sprung mass on the front left semicorner is half of the sprung mass of the corresponding axle and can be calculated with following formula:

$$M_{sfl}[kg] = \frac{1}{2} M_s \frac{b}{p}$$

Similarly, the mass present on the left rear semicorner is instead:

$$M_{srl}[kg] = \frac{1}{2} M_s \frac{a}{p}$$

With a, b e p respectively front half – wheelbase (distance between center of gravity and front axle), rear half – wheelbase and wheelbase of the vehicle.
The consequent force F_g that the front sprung mass exerts onto the ground is equal to:

$$F_g [N] = 9.81\, M_{sfl}$$

This force must be balanced by the preload of the front left spring which must support the front left sprung mass. To do this the spring must have the following preload in Newton:

$$F_s [N] = F_g \frac{1}{Lambda}$$

Given the bump frequency (or rebound frequency) f in Hz, the ground stiffness of the suspension follows the formula:

$$K_{susp}[N/mm] = \frac{1}{1000} (2f\pi)^2\, M_{sfl}$$

and being:

$$K_{spring}[N/mm] = K_{susp} \frac{1}{Lambda^2}$$

by replacing:

$$K_{spring}[N/mm] = (2f\pi)^2 \, M_{sfl} \, \frac{1}{Lambda^2} \, \frac{1}{1000}$$

We shall now also consider the tire stiffness, according to the scheme of the figure 4. Since the spring stiffness brought to ground and the stiffness of the tire are in series, the resultant stiffness of the suspension, which we will call *Ksusp,tire*, becomes:

$$K_{susp,tire}[N/mm] = \frac{K_{susp} K_t}{K_{susp} + K_t}$$

From which it can be seen that, if the tire stiffness is very high, in the denominator *Ksusp* can be neglected, obtaining:

$$K_{susp,tire}[N/mm] = \frac{K_{susp} K_t}{K_t}$$

which by simplifying becomes:

$$K_{susp,tire}[N/mm] = K_{susp}$$

In which we remember that the *Ksusp* is the stiffness of the spring brought to ground according the known relationship:

$$K_{susp}[N/mm] = K_{spring} Lambda^2$$

Generally, the spring stiffness to ground is in the order of tens of N/mm while the tire stiffness is in the order of hundreds of N/mm and therefore no big mistake is made if the tire stiffness is neglected in the calculation of the suspension stiffness at ground.

The suspension stiffness *Ksusp*, which does not take into account the tire, is called *wheel rate*, while the suspension stiffness *Ksusp,tire*, which takes into account the tire, is called *wheel ride rate*.

Below is an example.

We shall now calculate the stiffness and the preload of the front spring of the car with the following characteristics:

   Total mass = 811 kg
   Front unsprung mass = 63 kg
   Masses distribution = 66.1% front
   Lambda front = 0.95

We shall then proceed with the formula:

$$M_{sfl}[kg] = \frac{1}{2} M 0.661 - \frac{1}{2} M_{ns} = 236.5 \, kg$$

Considering a bump frequency equal to 2.41 Hz indicated for a race car.

$$K_{susp} \text{ [N/mm]} = (2f\pi)^2 \, M_{s_{fl}} \frac{1}{1000} = (2 \times 2.41 \times \pi)^2 \times 236.5/1000 = 54.2 \text{ N/mm}$$

$$K_{spring} \text{ [N/mm]} = K_{susp} \frac{1}{Lambda^2} = 54.2 \times \frac{1}{0.95^2} = 60.1 \text{ N/mm}$$

The preload of the spring:

$$F_s \text{ [N]} = F_g \frac{1}{Lambda} = 236.5 \times 9.81 \times \frac{1}{0.95} = 2441.8 \text{ N}$$

Where the F_g is the weight force of the front left sprung mass and it is equal to the product of the mass and the gravity acceleration.

Now we need to identify the spring that responds to the calculated characteristics (stiffness and preload) and verify that the installation on the suspension is feasible. To proceed, it is necessary to make further considerations that will allow us to choose the spring with the dimensional characteristics consistent with the compression and the extension travel expected. We shall examine the case of figure 5 in which the spring has the same axis of the shock absorber. In this regard, we define the following quantities:

1. *Lfree*: free length of the spring. Quantity provided by the manufacturer.
2. *Linst*: installed length of the spring. It represents the length of the spring loaded with the preload *Fs*. Quantity that depends on the suspension.
3. *Lmin*: minimum length of the spring. The spring cannot work at shorter lengths otherwise the coils come into contact. Quantity provided by the manufacturer.
4. *Ccompr*: Compression travel of the spring. In this case (figure 5), it depends on the maximum compression travel of the shock absorber.
5. *Cext*: Extension travel of the spring. In this case (figure 5), it depends on the maximum extension travel of the shock absorber.
6. *AB*: Distance between the shock absorber attachment points when installed (figure 5). Quantity that depends on the suspension.
7. *ABmax*: Distance between the attachment points when the shock absorber is free and fully extended. Quantity provided by the manufacturer.
8. *ABmin*: Distance between the attachment points when the shock absorber is free and fully compressed. Quantity provided by the manufacturer.

Let's define the compression and extension travels of the suspension which, in this case, depend on the physical characteristics of the shock absorber. The expression of the total travel allowed by the shock absorber is:

$$\text{TOTAL TRAVEL} = ABmax - ABmin$$

The expressions of the extension and compression travels of the spring and shock absorber, which in the case examined are the same, are:

$$C_{ext} = AB_{max} - AB$$

$$C_{compr} = AB - AB_{min}$$

During suspension extension, the spring load must never be canceled, otherwise the spring will detach from its seats. This places an upper limit on the extension travel. In particular, it must be:

$$F_s - C_{ext} K_{spring} > 0$$

$$C_{ext} < \frac{F_s}{K_{spring}}$$

If this inequality is not verified, the spring detaches from its seats. Therefore, if the extension travel is known, it is required to do this check. By replacing the previous expression of C_{ext}, we obtain the dependence on the travel of the shock absorber:

$$AB_{max} < \frac{F_s}{K_{spring}} + AB$$

If the maximum extension of the shock absorber is too large, it must be lowered using an internal rebound stop.

In the phase of maximum compression of the shock absorber, the spring must never reach its minimum length, otherwise the coils will come into contact. This places a lower limit on the maximum compression travel. In particular, it must be:

$$L_{inst} - C_{compr} > L_{min}$$

e then

$$C_{compr} < L_{inst} - L_{min}$$

If this inequality is not verified, the coils of the spring come into contact before the maximum compression of the shock absorber. Therefore, if the compression travel is known, it is necessary to make this check and, if required, choose a spring with a lower L_{min}.

By replacing the previous expression of C_{compr}, we obtain the dependence on the travel of the shock absorber

$$AB_{min} > AB - L_{inst} + L_{min}$$

If the maximum compression travel of the shock absorber is too large, it must be lowered by inserting a rigid element between the strut housing of shock absorber and the upper plate or it is necessary to intervene on L_{min}.

3.6. The roll and yaw motions

Roll and yaw motions are the most important motions in the study of the vehicle lateral dynamics when cornering. The first is a relative motion of the sprung mass with respect to the unsprung masses, the second is an absolute motion of the entire mass of the vehicle. Roll motion is the rotation of the sprung mass approximately[1] around the roll axis. This rotation is generated when the sprung mass is pushed towards the outside of the cornering by the centrifugal transverse inertia force.

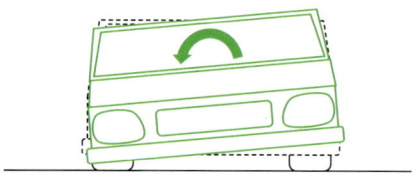

Figure 6 – Roll motion.

The position of the roll axis, as we will see later, depends on the architecture and geometry of the suspension. The roll stiffness depends on the stiffness of the springs but above all on the stiffness of the anti-roll bars. These are devices that give to each axle greater roll stiffness without varying the rebound and pitch stiffness which, as we have seen, are fundamental for the *comfort* of the vehicle.

The position of the roll axis and the stiffnesses greatly influence the vertical load transfers which are responsible for the slip angles taken by each wheel.

The velocity with which the vehicle rolls to position itself at a certain angle influences the trend over time of load transfers and the transverse forces between tire and road that balance the centrifugal force. The steering wheel angle which the user imposes in the cornering is not fixed but progressive and stabilizes when the vehicle reaches the final roll angle and the vertical loads stabilize.

The yaw motion is the absolute rotation of the entire vehicle around the vertical axis (z axis of figure). The faster this rotation is, the faster the vehicle turns.

The yaw motion overlaps to the roll motion because when the user applies the steering wheel angle the vehicle tends to both curve and roll at the same time. The motion becomes pure yaw when the vehicle stabilizes reaching the constant roll angle. In racing vehicles, it is important to limit this overlap by increasing the roll speed as much as possible through a good compromise between the springs, shock absorbers and anti-roll bars, in order to accelerate the stabilization of the vehicle when cornering and to make it faster.

---

[1]. The approximation lies in the fact that the shock absorbers and the tires alter the roll kinematics in the transient (see chapter 6). However, as will be shown later, a reference axis is required for sizing roll stiffnesses.

The motions of the vehicle

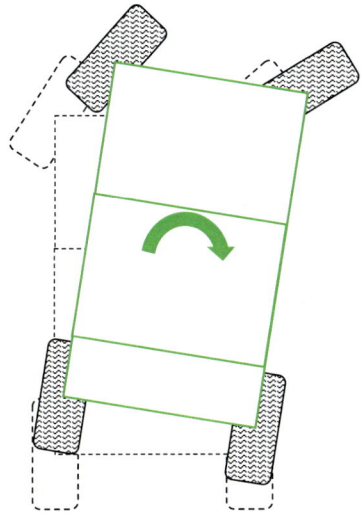

Figure 7 – Yaw motion.

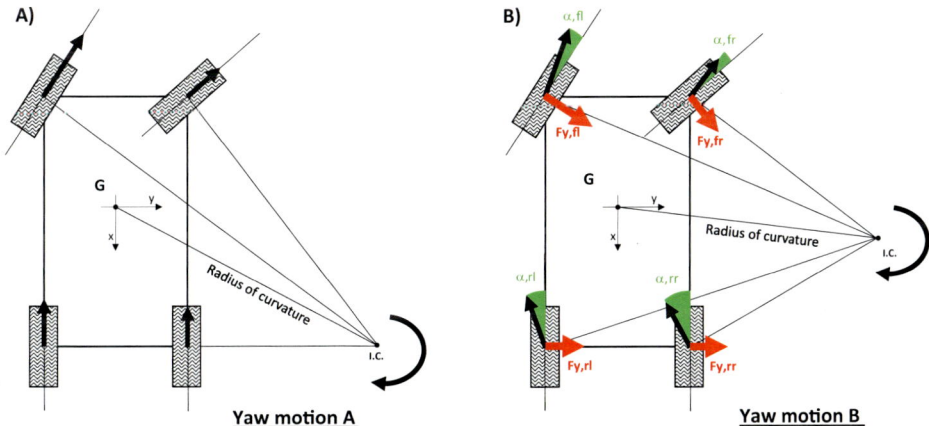

Figure 8 – Yaw motion with different slip angles, in steady state condition.

From the previous figure it is clear that, with the same vehicle and same steering wheel angle, the yaw motion depends on how the four wheels react and therefore on the slip angles of the tires. In fact, the vehicle A, which does not have slip angles, with the same steering wheel angle, assumes a greater radius of curvature. This means that, with the same initial speed, the vehicle B has higher yaw velocity. From this it is understood that, in order to obtain a certain velocity when cornering, it is necessary to plan well the relative reaction of the four wheels which, as we will see later, depends not only on the type of tire but also on the behavior of the suspensions.

Automotive suspension

3.7. The roll stiffness

The roll stiffness, evaluated in *Nmm/radians*, represents the ratio between the applied moment to the sprung mass and the corresponding rotation angle around the roll axis. This quantity depends on the combined reaction of the front axle and the rear axle. In addition to the sum of the reactions of the two axles, which define the total stiffness, it is important to evaluate their distribution because, as we will see in detail below, it is responsible for the repartition of load transfer between the axles. The roll stiffness of each axle depends on the springs and on the anti-roll bar which reacts torsionally only when the wheels move in opposite wheel travel.

With reference to the front axle we can measure the stiffness in opposite wheel travel $K_{opp,f}$ by making the ratio between the force *dF* and the displacement *dz* obtained by imposing, with the vehicle body locked, an opposite wheel travel to the two wheels.

$$K_{opp,f}[N/mm] = \frac{dF}{dz}$$

If t_f is the distance in mm between the centers of the contact patches (wheel track at ground), from the stiffness in opposite wheel travel we obtain the roll stiffness K_f of the axle by applying the following formula:

$$K_f[Nmm/radians] = K_{opp,f}\frac{t_f^2}{2}$$

In which, if the t_f,wc value of the wheel track at the wheel center is known, which is generally easier to find, t_f can be calculated using the following formula:

$$t_f [mm] = t_f,wc + 2RSC_f \tan(\gamma_f)$$

In which the *camber* is expressed in radians and *RSC* represents the radius under load in mm or, with a good approximation, the height from the ground of the wheel center. Similarly, for the rear axle:

$$K_r[Nmm/radians] = K_{opp,r}\frac{t_r^2}{2}$$

$$t_r [mm] = t_r, wc + 2RSC_r \tan(\gamma_r)$$

3.8. The sizing of anti-roll bar

The stiffness *Kopp,f* is the sum of the bump stiffness due to the spring and the stiffness in opposite wheel travel due to the anti-roll bar only which from now on we will call *Kbar,f*. It therefore turns out:

$$K_{opp,f}[\text{N/mm}] = K_{susp,f} + K_{bar,f}$$

To have a desired *Kbar* value it is necessary to act on the dimensions of the Anti-roll bar. We shall refer to the scheme of figure 10 which shows the quantities that define the operation of the bar. Generally, the dimensions *Lb* and *Lc* derive from the need to install the anti-roll bar without interfering with the various components of the suspension. Consequently, the diameter is the only variable in the sizing. In order to control the bar stiffness, the branch *Lc* should be very rigid in flexure so that its deformation does not overlap with the torsion of the central section. In this way the bar stiffness depends only on the diameter. As for the spring, for the bar sizing it is necessary to define the *Lambda,b* ratio between the displacement of the attachment point of the bar to the suspension and the displacement of the point at ground which results to be:

$$Lambda,b = \frac{S_b}{S_g}$$

In the following configuration the bar is anchored to the lower control arm while in other configuration it is anchored to the shock absorber through a rod with ball joints.

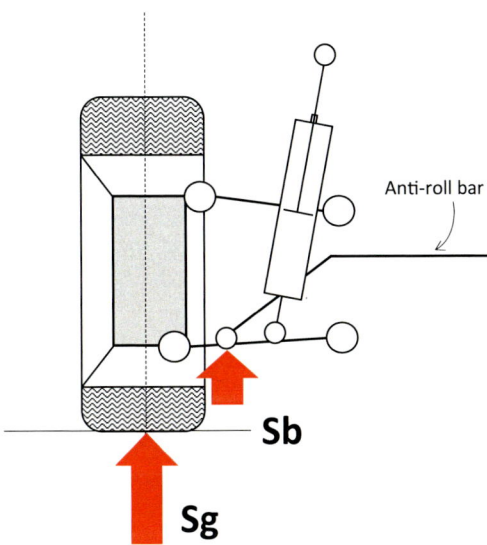

Figure 9 – Lambda,b ratio = Sb/Sg.

We shall now calculate *Kbar,susp* or the equivalent stiffness of the anti-roll bar (as if it

Automotive suspension

were a spring) at the anchor point on the suspension. By starting from the bar stiffness on the ground, which is generally the target to be achieved, through the *Lambda,b* ratio, we obtain *Kbar,susp* which will be used to evaluate the bar diameter. From what has been said it appears that:

$$K_{bar,f} [N/mm] = K_{bar,susp} (Lambda,b)^2$$

We shall then consider the bar schematized in the following figure and write the equation between the work of external forces and the work of internal forces. The two opposing forces *F*, which produce the torsion of the bar and the consequent angular deformation θ, represent the external stresses. The internal reaction moment, due to the torsional stiffness *Kb* of the bar, represents the internal stress.

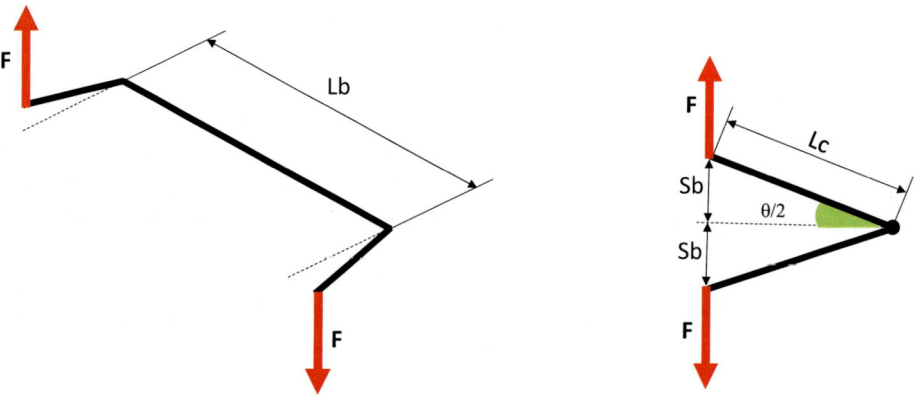

Figure 10 – Scheme of an anti-roll bar with the quantities necessary for sizing.

In the hypothesis of small displacements, it results:

$$2FS_b = \frac{1}{2} K_b \theta^2$$

$$F = \frac{K_b \theta^2}{4 S_b}$$

with:

$$S_b = L_c \frac{\theta}{2}$$

and then:

$$F = \frac{K_b \theta}{2 L_c}$$

The researched stiffness is:

$$K_{bar,susp} [N/mm] = \frac{F}{S_b} = \frac{K_b \theta}{2L_c} \frac{2}{L_c \theta}$$

$$K_{bar,susp} [N/mm] = \frac{K_b}{L_c^2}$$

The ground stiffness resulting from the anti-roll bar is as follows:

$$K_{bar,f} [N/mm] = K_{bar,susp} (Lambda, b)^2$$

$$K_{bar,f} [N/mm] = K_b \left(\frac{Lambda, b}{L_c} \right)^2$$

Where *Kb* is the torsional stiffness of the bar calculated as follows:

$$K_b [Nmm/radians] = G \frac{I_p}{L_b}$$

In which:
G, is the transverse modulus of elasticity of steel and is approximately 80233 N/mm²
Ip is the polar moment of inertia for circular hollow sections:

$$I_p [mm^2] = \frac{\pi [(Dext/2)^4 - (Dint/2)^4]}{2}$$

In which *Dext* e *Dint* are respectively the external diameter and the internal diameter in mm of the hollow section of the bar that allow to obtain the target value for *Kbar,f*. This formulation is correct if the branches *Lc* of the bar are very rigid and the bar has a straight path, valid hypotheses when making handcrafted bars for racing vehicles. When these hypotheses are not valid, it is necessary to use CAD software for modelling and FEM codes for sizing.

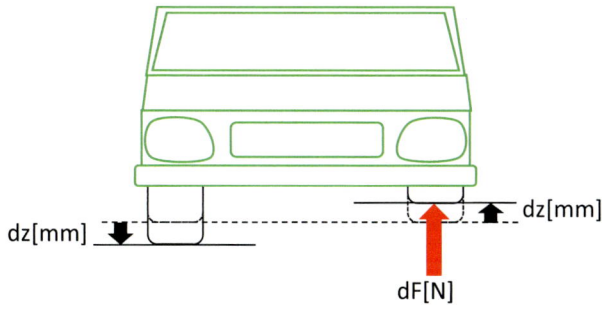

Figure 11 – Evaluation of the contribution of the bar in the roll stiffness of an axle.

A practical method to measure the contribution of the bar in opposite wheel travel, without having special equipment, is the following:

- Remove the springs;
- Lift the vehicle on a scissor lift leaving the wheels hanging;
- Lift the left wheel using a hoist until the configuration is the same as when the car is on the ground (referring to the distance between the wheel center and a point on the vehicle body);
- Hang a known mass M_p at the right wheel center and read the displacement dz in mm downwards.

The stiffness in opposite wheel travel will be:

$$K_{bar,f}[N/mm] = 2\,\frac{9.81 M_p}{dz}$$

Remembering that the displacement dz is relative to the position of the right wheel which is certainly lower than the left because it is influenced by the weight of the unsprung mass. It is advisable to measure the displacement starting from this configuration because, in this way, all the clearances of the bar joints are recovered, and the measurement is more precise.

This is a fairly approximate method but useful when an equipment capable of applying, with the vehicle body locked, an opposite wheel travel and detecting the emerging vertical forces is not available. There are also test benches that are able to evaluate these quantities more precisely, such as the *K&C* bench (*Kinematic and Compliance*).

3.9. The load transfer in stationary condition

When the vehicle travels along a curve of radius R, at constant velocity V, its center of gravity acquires a centripetal acceleration which generates a centrifugal force of inertia, which tends to push the car towards the outside of the curve. Let M be the mass of vehicle, the centrifugal force is:

$$F_c[N] = M\,\frac{V^2}{R}$$

The centrifugal force causes the vertical load transfer, from the inner wheels to the outer wheels, which increases with increasing centripetal acceleration. The condition in which the inner wheels lose all their load represents the rollover condition of the vehicle.

We shall consider the vehicle in the following figure on which the centrifugal force is applied.

The motions of the vehicle

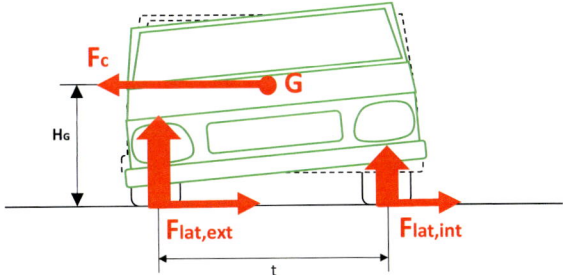

Figure 12 – Load transfer due to lateral centrifugal force.

The total load transfer is obtained from the equilibrium of the moment of the lateral force and the vertical forces:

$$\Delta Fz, total\,[N] = \frac{MA_{Gy}H_G}{t}$$

where t is the average of the wheel track at ground of the front and rear axle.

We can affirm that:
1. The total load transfer, namely the difference between the external vertical forces and the internal vertical forces of the curve:
   a. increases with the increase in the height of the center of gravity;
   b. decreases with the increase in the wheel tracks;
   c. does not depend on the geometry of the suspension and on the roll stiffness.
   With reference to the single axle, we can state that a car with a greater wheel track or a lower center of gravity, will have, with the same lateral force, less load transfer and consequently a lower slip angle on the relative axle (see chapter 2).
2. The sum of vertical forces is always equal to the weight force Mg because the outer side is loaded with the same amount the inner side is unloaded.
3. The lateral force on each tire is equal to vertical load multiplied by the adherence coefficient μ_y used.
4. In the case of $\mu_y = 1$ the sum of the lateral forces is equal to Mg. This is because each lateral force is equal to the vertical force which insists on the tire and since the sum of vertical forces is equal to Mg also the sum of the lateral force is equal to Mg.

From what has been said, we can state that, if μ_y is the maximum coefficient of adherence between the tire and the road, the maximum lateral acceleration is $g\mu_y$. This last statement does not take into account the tendency of the vehicle to roll over which increases as the ratio between half wheel track and center of gravity decreases. In fact, from the equilibrium of the lateral forces it can be seen that the maximum lateral acceleration obtainable is the minimum between that which causes the *roll over* and that which corresponds to the maximum adherence. In the simplified hypothesis in which the *roll over* occurs simultaneously on the two axles, we can evaluate the maximum lateral acceleration (at the limit of the *roll over*) by equating the total load transfer to the weight force present on one side of the vehicle which is equal to half of the total weight. It therefore follows that:

$$\Delta F_{z,total} = \frac{Mg}{2}$$

$$\frac{MA_{Gy}H_G}{t} = \frac{Mg}{2}$$

$$A_{Gy} = \frac{t}{2H_G}g$$

Combining with the above, the maximum lateral acceleration will be as follows:

$$A_{Gy,max} = min\left(\frac{t}{2H_G}g,\ g\mu_y\right)$$

From which it is clear that vehicles with high center of gravity or low tracks rolls over before reaching maximum adherence.

Generally, tires for road cars have μ_y approximately equal to 1-1.2 and therefore, except for *roll over*, the maximum lateral acceleration is 1.2g, while racing tires have decidedly higher adherence coefficients and, for this reason, they allow higher values of lateral acceleration. In formula 1 cars, the lateral acceleration values of around *5g* are achieved both thanks higher coefficients of adherence and thanks to a high vertical load resulting from the aerodynamic downforce. This is valid when the center of gravity is low enough not to cause the vehicle to roll over before side sliding. For example, if we equip a small car (H_G = 550 mm and t = 1340 mm) with tires at μ_y = 1.5, since the $t/(2H_G)$ ratio is 1.2, the vehicle rolls over before reaching maximum adherence. Therefore, in these cases, in addition to being useless it is also dangerous to adopt such high-performance tires because the crisis due to *roll over* is less welcome than that due to lateral sliding. The *roll over* considerations can be used indirectly to evaluate the height of the center of gravity if there is no other information to measure it (stabilized roll angle in the step steer or direct measurement with the chaotic pendulum). In fact, if the vehicle acquires a lateral acceleration of 1.5g without lifting the internal wheels, it can be said that the first member of the previous formula is greater than the second and therefore:

$$\frac{t}{2H_G} > 1.5$$

by entering the *t* of the previous car:

$$H_G < 447\ mm$$

From what we have seen, we can assert that the lateral acceleration corresponding to the vehicle roll over limit provides an important indication in evaluating the height of center of gravity. Similar considerations apply to the vertical load transfer due to the braking.

3.10. The distribution of load transfers in stationary conditions

In order to evaluate the load transfer, we schematize the vehicle with a sprung mass M_s which rotates around the roll axis thanks to the centrifugal force $M_s a_y$ and which has the roll stiffnesses of the two axles as impediments to rotation, as shown in figure below. To calculate the front and rear load transfer, we build an equivalent system by distributing the stresses acting on the center of gravity on the two axles.

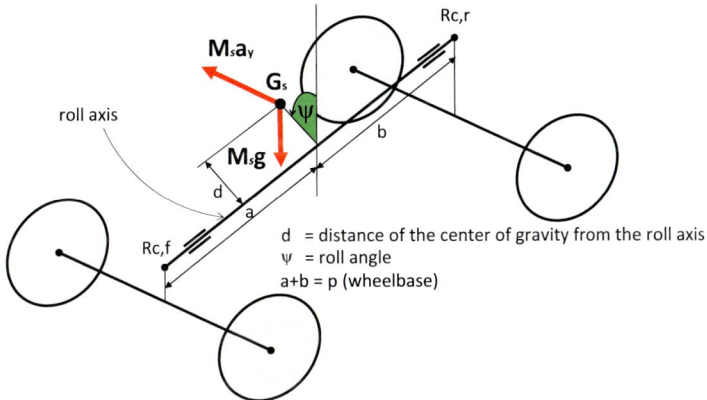

d = distance of the center of gravity from the roll axis
ψ = roll angle
a+b = p (wheelbase)

Figure 13 – Schematic of the vehicle with the sprung mass rotating around the roll axis. The sprung mass rolls by a ψ angle and the weight force moves laterally by $d \cdot tg(\psi)$.

As a first step, neglecting for the moment the weight force, we transport the lateral force $M_s a_y$ on the roll axis and add the torque $M_s a_y d$, called roll torque, where d is the distance of the center of gravity of M_s from the roll axis, as described in the following figure.

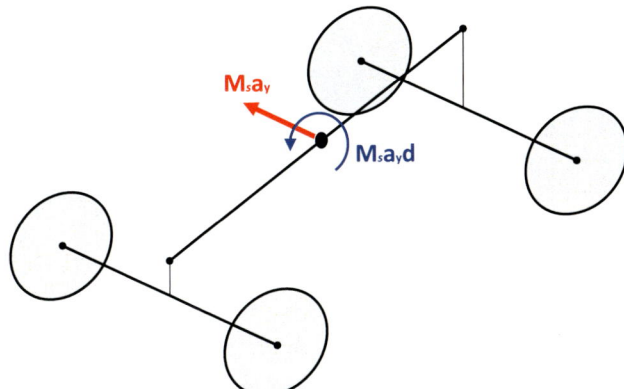

Figure 14 – First equivalent system obtained by transporting the centrifugal force on the roll axis and adding the transport torque.

Automotive suspension

The second step in building the equivalent system is to distribute the lateral force and the roll torque on the two axles. In the following figure the definitive equivalent system.

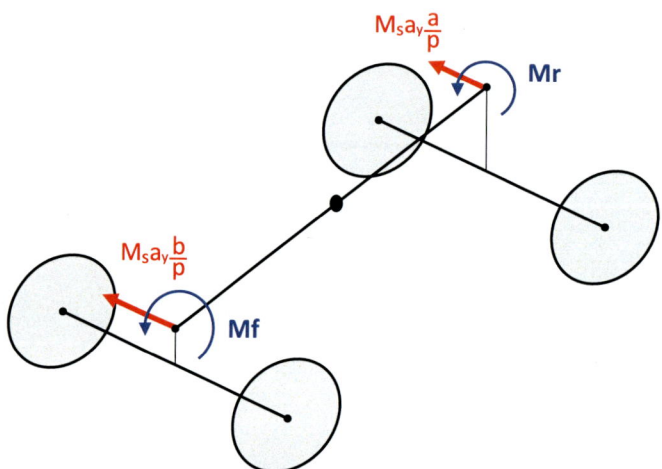

Figure 15 – Definitive equivalent system obtained by distributing lateral force and roll torque on the axles.

The transport torque $M_s a_y d$ is distributed on two axles following the law of *fixed ended beam*, by loading more the axle with the greater stiffness.

$$Mf\ [Nm] = \frac{Kf}{Kf + Kr} M_s a_y d$$

$$Mr\ [Nm] = \frac{Kr}{Kf + Kr} M_s a_y d$$

With *Kf* and *Kr* respectively the roll stiffness of the two axles.
The lateral force is distributed on the two axles following the criterion of *the simply supported beam*, by loading more the axle with the greater vertical weight:

$$Ff\ [N] = \frac{b}{a + b} M_s a_y$$

$$Fr\ [N] = \frac{a}{a + b} M_s a_y$$

The following components are added on each axle to the above forces and torques, due to the sprung mass:

· A lateral force due to the centripetal acceleration of the unsprung mass which, without making major errors, is positioned at the center of the wheel at a height equal

The motions of the vehicle

to the radius under load.
· A torque that depends on the lateral displacement of the center of gravity due to the roll. In the following figure, the front axle with all the forces and torques that produce the load transfer is shown.

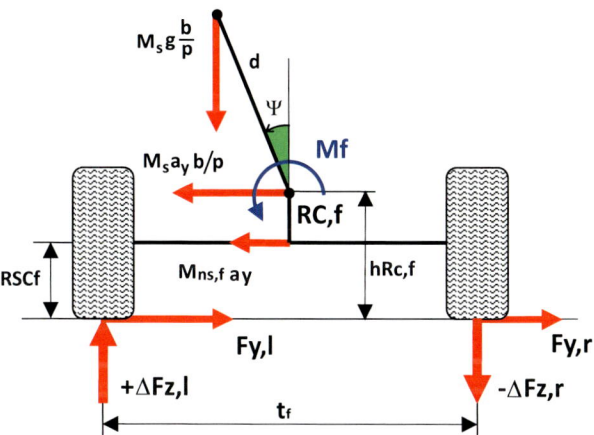

Figure 16 – Forces and torques acting on the front axle when cornering.

The front axle load transfer will therefore have the following four contributions:

1. The contribution due to the roll torque M_f:

$$\Delta F_{z,roll,f} = \frac{K_f}{K_f + K_r} M_s a_y d \frac{1}{t_f}$$

This contribution becomes influential when the distance d between the center of gravity and the roll axis increases. In this case, by modulating the roll stiffnesses K_f and K_r, the load transfers and consequently the slip angles of the two axles can be modulated. As the distance d increases, it is necessary to adopt stiffer bars to contain the roll angle.

2. The contribution due to lateral force:

$$\Delta F_{z,lateral,f} = M_s a_y \frac{b}{p} \frac{hR_{c,f}}{t_f}$$

This contribution becomes influential when the height of the roll center increases, which corresponds to the decrease in the distance d. Generally, as contribution 2 increases, contribution 1 decreases and the load transfer follows only the weight distribution, thus losing the ability to modulate it on the two axles by adjusting the roll stiffnesses.

3. The contribution due to the unsprung mass:

$$\Delta F_{z,Mns,f} = M_{ns,f} a_y \frac{RSC_f}{t_f}$$

with RSC_f equal to front unloaded radius. This contribution cannot be modulated.

4. The contribution due to the transverse displacement of the weight force:

$$\Delta F_{z,wf,f} = M_s g \frac{b}{p} \frac{d \tan\psi}{t_f}$$

This contribution cannot be modulated.
On the rear axle the situation is similar.
By adding the four contributions we obtain the total load transfer on the front and on the rear axle.

$$\Delta F_{z,f} = \frac{M_s a_y}{t_f} \left(\frac{K_f}{K_f + K_r} d + \frac{b}{p} hR_{c,f} \right) + M_s g \frac{b}{p} \frac{d \tan\psi}{t_f} + M_{ns,f} a_y \frac{RSC_f}{t_f}$$

$$\Delta F_{z,r} = \frac{M_s a_y}{t_r} \left(\frac{K_r}{K_f + K_r} d + \frac{a}{p} hR_{c,r} \right) + M_s g \frac{a}{p} \frac{d \tan\psi}{t_r} + M_{ns,r} a_y \frac{RSC_r}{t_r}$$

The contributions we have called $\Delta F_{z,lateral}$ and $\Delta F_{z,Mns}$ are certainly faster than the others because they appear as soon as the lateral acceleration arises. Instead, the contributions $\Delta F_{z,roll}$ and $\Delta F_{z,wf}$ have a certain delay compared to the previous ones. This is because the first requires the roll stiffness of the axles to intervene, while the second requires the vehicle body roll which moves the weight force laterally. This follows from the stiffnesses not reacting immediately, due to the presence of the clearances and elastic elements and the roll being delayed by the damping of the shock absorbers. In order to speed up the vehicle response when cornering, the distance d could be reduced as much as possible. In this way, the weight of contributions 2 and 3 would be increased and the moment of inertia around the roll axis would be lowered, with the consequent acceleration around it. The limit lies on the fact that, if d is too small, contribution 1 loses influence with respect contribution 2 and the possibility of modulating the load transfers through the anti-roll bar is lost. This becomes a problem when it is necessary to modify the distributions of the slip angles of the axles by acting on the roll stiffness. This is the case of a rear-wheel drive vehicle and rearward engine that, in stationary conditions, has a decidedly oversteering tendency because the arrangement of the masses moves greater lateral force on the rear axle which already tends to derive more than the front axle due to the traction. If the roll axis is too close to the center of gravity and the contribution 1 loses importance, the vehicle certainly has a rapid roll response but has less possibility of moving the distribution of the load transfers towards the front axle in order to increase its slip and attenuate its natural tendency to oversteer.
Sometimes it is decided to decrease the distance d by raising the roll axis in order to reduce the diameter of the bars which, if too rigid, do not allow the complete extension of the suspension, favoring the detachment of the inner wheel at the curve. This technique is widely used in off-road vehicles, where it is desirable that the wheels never lose contact with the road. The height limit of the roll axis derives from the fact that, when cor-

The motions of the vehicle

nering, on each of the two wheels generates a moment *M,roll*, which causes the vehicle body to roll, and a moment *M,lift*, which causes it to lift as shown in the following figure:

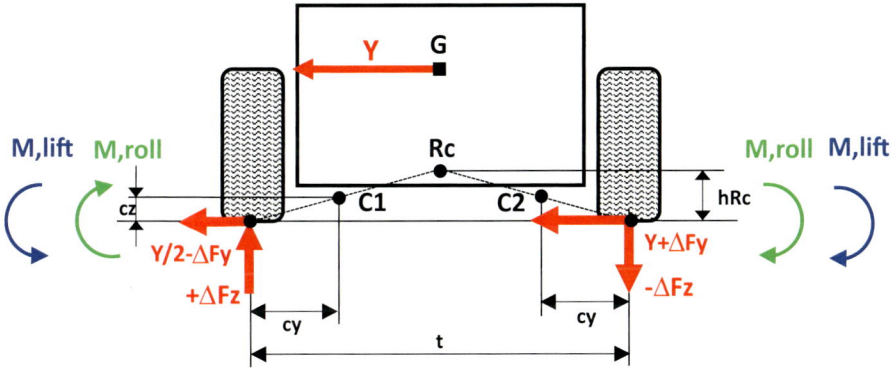

Figure 17 – Arrangement of the roll center and instantaneous centers with respect to the forces in equilibrium.

Where *C1* and *C2* are the position of the instantaneous centers of the suspension kinematics projected on the transverse plane. They are called *ICFV (Instant center front view)* and their construction will be clearer in the chapter 6.
In the following two figures the details of the vehicle body roll and lift due respectively to the moments *M,roll* and *M,lift* are shown.

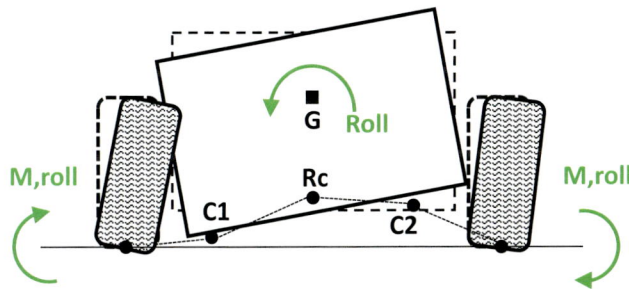

Figure 18 – Roll due to the two moments *M,roll*.

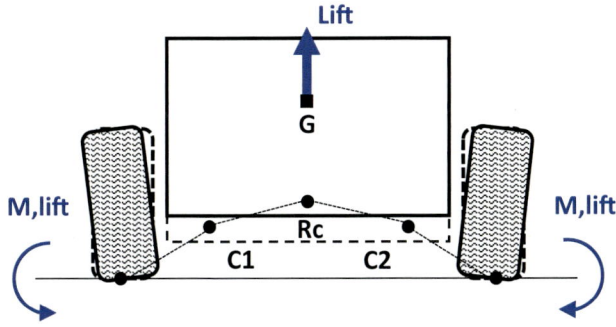

Figure 19 – Lift due to the two moments *M,lift*.

It can be demonstrated that the moments are:

$$M,\text{roll} = \Delta F_z cy - \frac{Y}{2} cz$$

$$M,\text{lift} = \Delta F_y cz$$

Where *cy* and *cz* determine the positions of the centers *C1* and *C2*. In particular we see the *M,lift* proportional to their height.

The same equations can be rewritten as a function of the height of the roll center.

$$M,\text{roll} = cy \left(\Delta F_z - Y \frac{hRc}{t} \right)$$

$$M,\text{lift} = cy\, \Delta F_y \frac{2hRc}{t}$$

In which it is clear that the torque *M,lift* increases as the height of the roll center and the distance cy increase. This is an undesirable behavior because the consequent lifting of the vehicle body produces several unwelcome behaviors such as lateral sliding, loss of camber, etc. The phenomenon of lifting due to *M,lift* is called *suspension jacking*.

In designing the height of the roll axis we must also keep in mind that the slip angle response of the axle occurs in two stages. The first stage is the one in which the slip angle due to the faster contributions (2 and 3) emerges, while the second stage is the one triggered by the roll. It is therefore necessary that the fastest contributions are well balanced between the axles so that the beginning of the second interval is common. It is equally important that there is not too much difference between the rapid contributions and the less rapid ones, so as not to differentiate too much the response times of the entire vehicle, and for this reason it is necessary to further modulate the height of the roll axis with respect to the position of the center of gravity.

Furthermore, we observe that the *DFz,lateral* (contribution 2) is proportional to both the weight and the roll center height of the axle and generally this contribution is greater on the most loaded axle. In a vehicle with greater weight distribution on the front axle, if the roll axis is inclined so that the rear roll center is much higher than the front one, it can be achieved that the *DFz,lateral, rear > DFz,lateral, front*. This condition, to be avoided because the slip angle of the less loaded axle increases (see chapter 4), is induced by an important inclination of the roll axis and occurs when:

$$\frac{hRc,r}{hRc,f} > \frac{b}{a} \frac{t_f}{t_r}$$

3.11. The roll angle

The calculation of the roll angle in stationary conditions has three main contributions.

Contribution 1. It is due to the rotation of the sprung mass around the roll axis and is:

$$\psi \text{ [deg]} = \frac{M_s a_y}{K_f + K_r} d$$

dividing by the lateral acceleration expressed in multiples of g (ay/9.81) we get the roll angle for each lateral g also called the roll gradient.

$$KROLL \text{ [deg/g]} = \frac{9.81 M_s}{K_f + K_r} d$$

Contribution 2. It is due to the lateral displacement of the sprung mass which generates an additional roll moment equal to the relative weight force $M_s g$ for the arm $d \tan\psi$.

Contribution 3. Due to vertical deformation of the tires.

Contributions 2 and 3 are considered only in the most accurate analysis.
The roll gradient is a measure of how much a car rolls for a lateral acceleration of *1g*. Generally, the road cars have roll gradients of about 6-7 *deg/g*, the sports cars about 3-5 *deg/g* while in the racing cars go down to 1-2 *deg/g*.

3.12. The sizing of the roll stiffness

From the analysis of the formulation of load vertical transfers emerges that a certainly important contribution is given by the term *ΔFz,roll* which depends on the distributions of the roll stiffnesses. Since the transfer of vertical load influences the slip angle assumed by an axle, as seen in chapter 2, it follows that, if d is large enough, the correct distribution of the roll stiffnesses is fundamental for the lateral vehicle dynamics. In particular, the slip angle increases as the load transfer increases which corresponds to the increase in the roll stiffness of the axle itself. Otherwise, if the roll axis is very close to the center of gravity, the *ΔFz,roll* contribution loses weight in the load transfer and the roll stiffnesses lose influence but, as already explained, this is a condition that should not occur.

The distribution of roll stiffnesses tends generally to follow that of the weight. For front-wheel driver road cars the percentage of the front stiffnesses distribution exceeds that of the weights by about 2-4%, for rear-wheel driver sports cars does not exceed 2% instead for slalom racing cars it can go also from -2% to -11% (more shifted towards rear axle).

In addition to the distribution, it is necessary to control the sum of the stiffnesses because the roll gradient depends on this.

Therefore, the operative steps in the stiffnesses sizing are the following:

1. Once the target roll gradient has been defined, by means of the position of the center of gravity and therefore of the distance *d* of the previous formula, the sum *Kf* + *Kr* = *Ktot* is calculated.
2. Once the target distribution has been defined, which depends on the type of vehicle, *Ka* and consequently *Kr* are calculated.
3. Once the stiffnesses of the two axles are identified, the anti-roll bars are sized so that the ground stiffness of the bar is the difference between the total stiffness of the axle and the ground stiffness of the spring.

We shall examine the example of one car with a front weight distribution equal to 65% and the target for the distribution of stiffnesses equal to +2%.

If the roll gradient is 6 degrees for each lateral g, we calculate *Kf+Kr* as follows:

$$(Kf + Kr) = \frac{M_s 9.81}{6} d$$

We calculate *Kf* and the consequent *Kr* by the equation of the stiffnesses distribution:

$$Kf = (0.65 + 0.02)(Kf + Kr) = 0.67 \, Kot$$

$$Kr = Kot - Kf$$

Being

$$Kf = K_{opp,f} \frac{t_f^2}{2}$$

We get

$$K_{opp,f} = Kf \frac{2}{t_f^2}$$

Being

$$K_{opp,f} = K_{susp,f} + K_{bar,f}$$

Knowing the ground stiffness due to the spring, which depends on the bump frequency, we can go back to the ground stiffness due to the bar alone and size its diameter as described in the previous paragraphs.

$$K_{bar,f} = Kf \frac{2}{t_f^2} - K_{susp,f}$$

$$K_{bar,r} = Kr \frac{2}{t_r^2} - K_{susp,r}$$

The following figure shows a spreadsheet with a complete example of a segment b vehicle in track set-up. The cells in green represent the input data, the others the calculated

The motions of the vehicle

values.

We start from the external dimensions, from the masses and from the ground frequency targets and we calculate the stiffness of the springs, the roll stiffnesses and the bar effects (*Kbar,f and Kbar,r*). In the final part of the calculation sheet there are the roll gradient, the three contributions to the load transfer and the loads of the wheels inside the curve for a lateral acceleration of *1g*. When the load of the inner wheel becomes small, there is the danger that the wheel will lift off the ground.

The reader can practice trying to calculate the same outputs.

Dimensions	
p [mm]	2407.0
t_f [mm]	1351.0
t_r [mm]	1329.0
Masses	
M [kg]	**840**
Percentage of front mass [%]	64.2
H_G [mm]	550
Mns,f [kg]	63
Mns,r [kg]	40
Ms [kg]	737
H_{Gs} (sprung mass) [m]	0.588
a [mm]	0.862
b [mm]	1.545
d [mm]	0.516
Tire	
RSCf [mm]	275.0
RSCr [mm]	278.0

Suspensions	
Front wheel track var. [mm/mm]	0.10
Rear wheel track var. [mm/mm]	0.12
hRc,f [mm]	68
hRc,r [mm]	80
Roll axis inclination [deg]	0.290
Frequency front. [Hz]	2.00
Frequency rear [Hz]	1.90
Lambda,f	0.95
Lambda,r	1.00
Ksusp,f [N/mm]	**37.6**
Ksusp,r [N/mm]	**18.6**
Kspring,f [N/mm]	**41.7**
Kspring,r [N/mm]	**18.6**
Kopp,f [N/mm]	**61.0**
Kopp,r [N/mm]	**32.2**
Sum Kopp [N/mm]	**93.2**
Kbar,f [N/mm]	**23.4**
Kbar,r [N/mm]	**13.6**

Roll - stiffness and gradient	
Ktot [Nm/deg]	1467.8
Kf [Nm/rad]	55668.6
Kr [Nm/rad]	28436.5
Roll gradient (contribution 1)	2.54
Roll gradient (contribution 2)	0.11
Roll gradient tot. [deg/g]	**2.7**
Percentage of roll stifness [%]	66.2

Load Transfer	
ΔFZ, front TOTAL [N/g]	2187
Load of internal wheel, front [N]	458
ΔZ,front [mm/g]	36
ΔFZ,roll,f [N/g]	**1829**
ΔFZ,lateral,f [N/g]	**232**
ΔFZ,Mns,f [N/g]	**125.8**
ΔFZ,rear TOTAL [N/g]	1187
Load of internal wheel, rear [N]	288
ΔZ,rear [mm/g]	37
ΔFZ,roll,r [N/g]	**950**
ΔFZ,lateral,r [N/g]	**155**
ΔFZ,Mns,r [N/g]	**82.1**

Figure 20 – Numerical values of the main characteristics controlled in lateral dynamics on a segment b car in track set-up.

In the calculation there are also the values of bump travel at ground (*Sg*) of the wheels outside the curve at lateral acceleration of 1g. They are noted as *ΔZ,front* and *ΔZ,rear* refer to the front and rear wheel respectively. It is important to predict these displacements to monitor the intervention of the suspension bump stops. For lateral acceleration that occur when cornering, it is advisable that the bump stops do not intervene because they produce undesirable increases in ground stiffness, with the consequent increase in load transfer and slip angle of the corresponding axle. In the left part of the following figure, the trend of the load-displacement curve during the bump travel of the suspension in which the increase in stiffness at ground after the intervention of the bump stop is evident. On the right side is typical installation scheme of the bump stop positioned on the strut housing of the shock absorber.

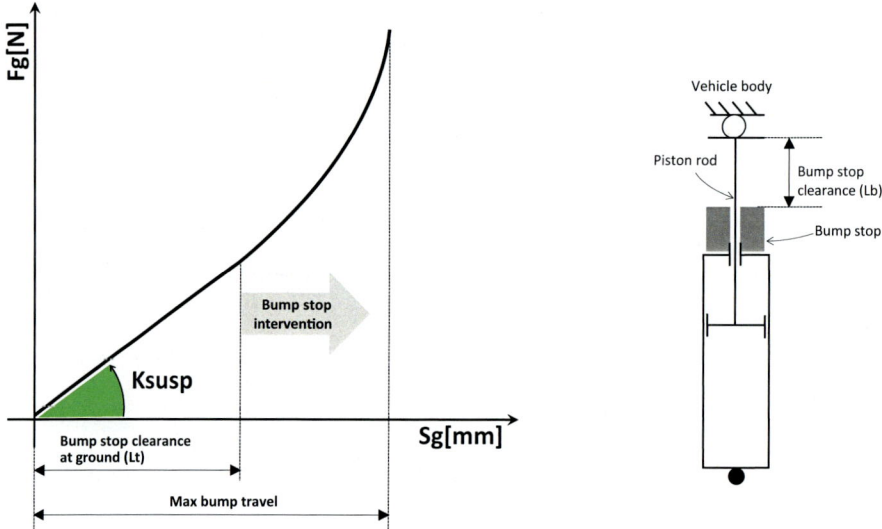

Figure 21 – Trend of the load-displacement curve of the semicorner evaluated on the ground. Typical bump stop installation scheme.

Knowing the ratio between the bump stop clearance L_b and the bump stop clearance at ground L_g:

$$Lambda, bump = \frac{L_b}{L_g}$$

in this case, typical of the McPherson and Double wishbone schemes, the ratio between the bump stop clearance L_b and the bump stop clearance at ground L_t is equal to the Lambda ratio of the spring. Having identified the maximum ground displacement S_g when cornering, on either axle it must be:

$$S_g Lambda, bump < L_b$$

If it is not possible to satisfy the relationship in all maneuvers, it is preferable that the front bump stop intervenes first in order to produce more understeer.

3.13. The sizing of the shock absorbers

This paragraph will provide guidelines for the rough sizing of the damping characteristics of the shock absorbers only with regard to handling performance.

The shock absorbers are dynamic dampers that serve to absorb kinetic energy, namely to stop or slow down the movement of objects when it becomes unwanted. For example, when the car faces a bump, the upward movement of the wheel is a desired movement because the consequent compression of the spring serves to distribute the force that reaches the vehicle body over time. Contrarily, the resulting vibratory movement is undesirable because it causes annoying accelerations to the passengers and the shock absorbers have the task of damping it in the shortest possible time.

Similarly, when making a step steer at a constant speed, the resulting roll is a desired movement because it serves to make the cornering comfortable; on the other hand, the vibratory roll movement prior to the stabilized condition is not desired and it is the duty of the shock absorbers to dampen this vibration. Curve 2 in the figure represents the desired trend of the roll angle in the face of a step steer maneuver at constant speed while curve 1 represents the undesired trend because the vibration produces vertical movements of the wheels that induce the user to make subsequent corrections of the steering wheel angle, as will become clear later.

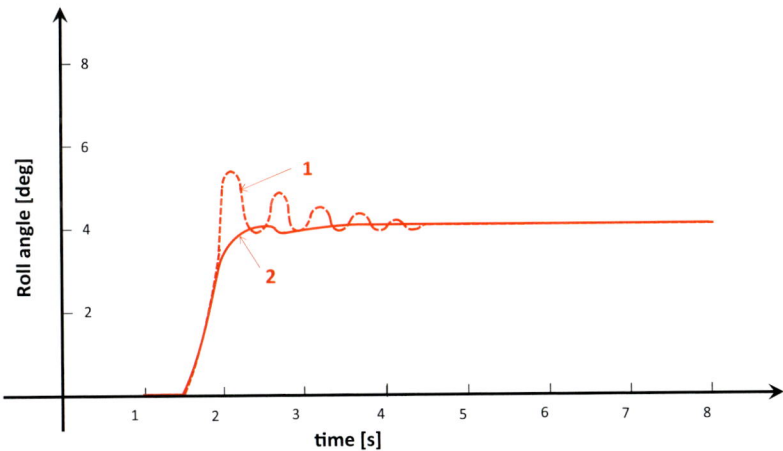

Figure 22 – Trend of the roll angle in a step steer maneuver at constant speed on a car with a different calibration of the shock absorbers.

When evaluating shock absorbers, we must distinguish between compression operation (bump) and extension operation (rebound). Unlike the spring, the shock absorber has an asymmetrical characteristic curve because the compression trend is always softer than the extension trend. The reason lies in the fact that the characteristic in compression, which dampens the vibrations deriving from the road roughness and lowers the accelerations transmitted to the vehicle body as much as possible, if too high increases the maximum values of exchanged forces with the consequent worsening of *comfort*. On the other hand, the extension characteristic, which is more important for handling, must be more energetic because it serves to dampen the vibrations resulting from the roll and it is not responsible for impulsive load transfers between the suspensions and the vehicle body. We shall follow with an example of damping characteristic of a road car in which the trend of the force exerted by shock absorber is represented as function of the speed of the piston rod with the evidence of the ranges of velocity relating to the different types of maneuvers. Indicatively, the 50-200 mm/s range is the one that contains the relative speeds (between the strut housing of the shock absorber and the piston rod) that are generally found in handling maneuvers, the 250-600 mm/s range concerns *comfort* maneuvers such as driving on the pavè, the obstacle passing and the long wave at high speed. This last maneuver consists in making the vehicle run, at a speed of at least 120 km/h, on a track where there is a long-wave depression to monitor the trend of the pitch angle which must have a peak-to-peak value content and an absolutely non-vibratory behavior.

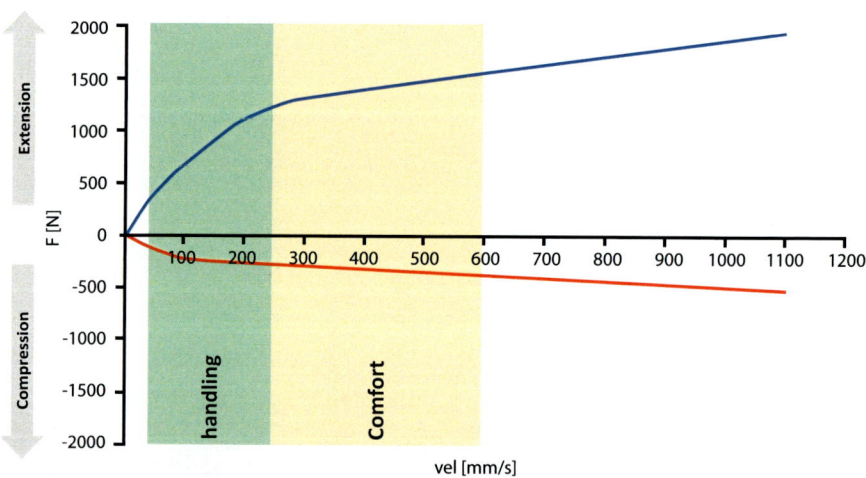

Figure 23 – Characteristic of a shock absorber for a road car with evidence of the speed ranges relative to handling and *comfort*.

For shock absorber sizing, we shall refer to the scheme of the following figure with the sprung mass and the unsprung mass of the semicorner. Here *Ksusp* and *C* are respectively the stiffness of the suspension and the damping of the shock absorber at ground, while *Kt* represents the stiffness of the tire. We provide the results and the relative observations directly regardless of the analytical solution of the system of differential equations, in order to focus on the correct application of the formula.

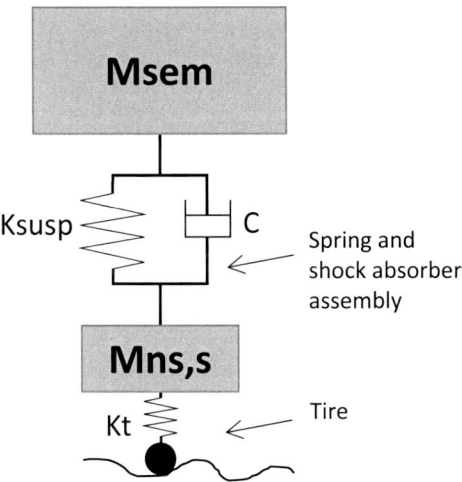

Figure 24 – Schematic of a semicorner with a two degrees of freedom model with two stiffnesses and a damping that represents the shock absorber at ground.

From the solution of the system of differential equations of the two-degree-of-freedom model in the figure, a *Copt* damping value is obtained, related to the compression characteristic, which minimizes the trend of the vertical accelerations of the sprung mass at different vibration frequencies. This condition is appreciated to ensure a good level of vehicle *comfort*. The formulation of *Copt* is as follows:

$$Copt = \sqrt{\frac{M_{sem} K_{susp}}{2}} \sqrt{\frac{K_t + 2K_{susp}}{K_t}}$$

where the *Copt* value represents damping at ground. In the event that the spring is coaxial with the shock absorber, the damping value related to it is:

$$Copt,shock = \frac{Copt}{Lambda^2}$$

The evaluation of the optimal compression damping depends on the vehicle mission. In the road vehicles aiming only to minimize vertical accelerations of the sprung mass, *Copt* is the ideal compression damping. From the analysis of the two degrees of freedom system, it can be seen that the damping that maximizes the trend of the load on the ground (on the tire), at the different vibration frequencies of the sprung mass, is greater than *Copt*. Therefore, in vehicles that aim to maximize dynamic performance, and therefore need a greater reserve of vertical load on the ground to face the demand for lateral force on the tires, it must be:

$$Cbump,shock > Copt,shock$$

and in particular:

$$C_{bump,shock} = q(C_{opt,shock})$$

where *q* varies from 1.2 to 2 depending on the level of sportiness of the vehicle. Obviously, the increase in *q* corresponds to the worsening of *comfort*.
Regarding the sizing of the extension characteristic, the value to be referred to is the critical damping of the sprung mass which has the following formulation:

$$C_{crit} = 2\sqrt{M_{sem}K_{susp}}$$

which becomes on the shock absorber

$$C_{crit,shock} = \frac{C_{crit}}{Lambda^2}$$

Generally, the starting damping for the extension characteristic is:

$$C_{ext,shock} = 0.8\,(C_{crit,shock})$$

The damping identified for the compression and extension characteristics are only the starting points because the final optimized values are the results of dynamics tests in a virtual environment or in an experimental environment. Furthermore, they represent only the slopes of the curves which must be suitably connected.
Below the calculations carried out to identify the initial slopes of the damping curves of the front shock absorber of the vehicle aforementioned.

Front shock absorber	
Msem [kg]	238.1
Mns,s [kg]	31.5
Kspring [N/mm]	41.7
Kt [N/mm]	400.0
freq [Hz]	2.0
Ksusp [N/mm]	37.6
Lambda	0.95
Copt [Ns/m]	2306.4
Ccrit [Ns/m]	5985.1
Copt,shock [Ns,m]	**2555.6**
Ccrit,shock [Ns,m]	**6631.7**
Cbump,shock [Ns/m]	5111.2
Cext,shock [Ns/m]	5305.4

Figure 25 – Calculations carried out for the sizing of the slopes of the compression and extension characteristics of the front shock absorber. The masses are relative to semicorner.

The slopes of the compression and extension characteristics of the shock absorber were calculated as follows:

$$Cbump, shock = 2(Copt, shock)$$

$$Cext, shock = 0.8(Ccrit, shock)$$

3.14. The dynamic influence of the shock absorbers

The shock absorbers influence the trend of the positions assumed by the wheel during the transients of the different maneuvers and have no weight when the vehicle has reached the steady state condition. When the damping is high, the shock absorbers suppress the relative degree of freedom of the kinematics, which in the first part of the transient behaves as if it were completely locked. If, for example, we image to increase the extension damping, in the first part of the pitch motion induced by braking, the rear suspension kinematics is locked and the vehicle does not rotate around the pitch axis, as in the left part of the following figure, but around a transverse axis very close to the centers of the rear contact patches. This phenomenon, in the first part of the transients, produces the displacement of the wheel (compression) only on the front suspension, with the behavioral consequences deriving from the influence of the new configuration assumed by the front wheels.

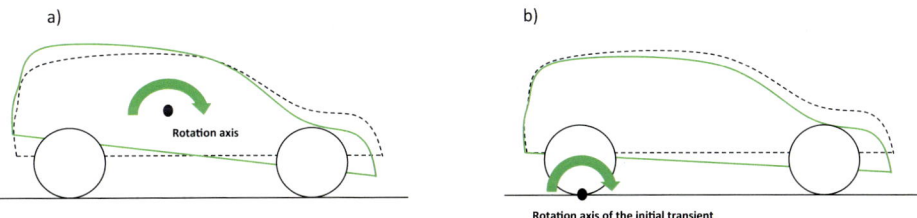

Figure 26 – a) Representation of the initial pitch transient obtained with shock absorbers with an almost symmetrical (compression-extension) characteristic. b) Representation of the initial pitch transient obtained by increasing the extension characteristic of the shock absorbers. Figure a) also represents the final stationary configuration of the pitch which does not depend on the shock absorbers.

The same happens with regard to the roll in which, during the first part of the transient, the axle rotates around the center of the contact patch of the wheel inside the curve. In this situation, the inner suspension remains locked without extending while the outer suspension compresses.

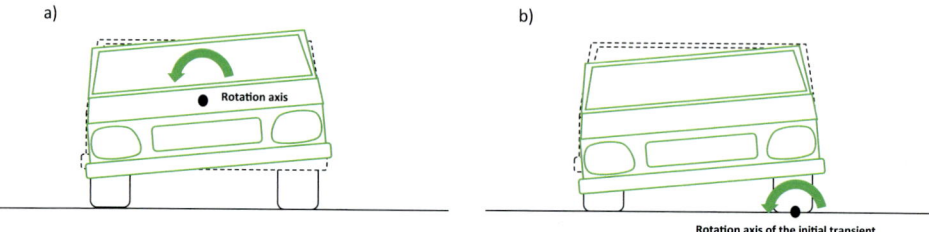

Figure 27 – a) Representation of the initial roll transient obtained with symmetrical (compression-rebound) shock absorbers. b) Representation of the initial roll transient obtained by increasing the extension characteristic of the shock absorbers. Figure a) also represents the final stationary roll configuration which does not depend on the shock absorber.

3.15. Summary

The rebound and pitch motions are important for the *comfort* of the vehicle and the oscillation frequency depends on the stiffness of the springs. The springs must never be too rigid because they would trigger annoying vibrations, even on sports vehicles. The designers should not make the mistake of stiffening the springs too much to lower the roll. Roll and yaw motions are important for vehicle *handling*.

The load transfers due to the lateral centrifugal force influence the distribution of the slip angles between the axles which, as we will see in the next chapters, define the understeer and oversteer tendency of the vehicle. The load transfer of the generic axle consists of 4 contributions:

1. The contribution due to the roll torque which depends on the distance *d* between the center of gravity and the roll axis and on the distribution of the roll stiffnesses. These, in turn, depend on the geometry of the anti-roll bars. In particular, when *d* is suitably large, it is possible to modulate the load transfers between the axles through the roll stiffnesses.

$$\Delta F_{z,roll,f} = \frac{K_f}{K_f + K_r} M_s a_y d \frac{1}{t_f}$$

2. The contribution due to the lateral force on the front axle, which depends on the weight distribution and the height of the roll center. It is important to size the position of the roll axis well so it is not placed too close to the center of gravity, because in this case the contribution 1, which offers excellent possibilities for tuning, would lose influence. Contribution 2 is less delayed than the previous one.

$$\Delta F_{z,lateral,f} = M_s a_y \frac{b}{p} \frac{h_{Rc,f}}{t_f}$$

3. The contribution due to unsprung masses:

$$\Delta F_{z,Mns,f} = M_{ns,f} a_y \frac{RSC_f}{t_f}$$

4. The contribution due to the transverse displacement of the weight force which reaches its maximum only at the end of the roll phase.

$$\Delta F_{z,wf,f} = M_s g \frac{b}{p} \frac{d \tan\psi}{t_f}$$

When contribution 2 does not have much influence, the distribution of the load transfer depends on the distribution of the roll stiffnesses between the two axles, which has a similar value to the weight distribution with a few more or less points depending on the vehicle mission. Starting from a correct distribution, the designer can increase the rear stiffness or decrease the front, to have more load transfer to the rear in order to increase the rear slip angle. This adjustment possibility is guaranteed only if contribution 1 has a certain influence in the global load transfer, that is when the roll axis is not too close to the center of gravity.

The next step is the sizing of the shock absorbers with the following formula

$$C_{bump,shock} = q(C_{opt,shock})$$

$$C_{ext,shock} = 0.8(C_{crit,shock})$$

which must be optimized by eliminating, firstly, the vibratory behavior of the roll angle trend. The measurement of the roll angle and lateral acceleration can be carried out by applying a capacitive accelerometer and a gyroscope near the center of gravity.

The calculation of load transfers also serves to define the lateral and longitudinal forces acting on the wheels during vehicle operation, to identify the level of tire utilization, having the characteristic curves available.

4. LATERAL DYNAMICS
Use of analytical models

4.1. The analytical approach

In the study of vehicle dynamics the purely analytic approach does not always lead to concrete results because it is very long and complex. To obtain faster and less laborious evaluations it is advisable to use either the virtual approach (*multibody*) or the experimental approach. In any case, analytical knowledge is essential to have both a first interpretation of the physical phenomenon and to control the virtual and the experimental approaches.

In this chapter we will try to simplify the analytical models, in order to make them more usable for the re-elaboration of the experimental measurements and to correctly interpret the physical phenomena.

Let's begin with the equivalent quadricycle. It consists, as already seen, in considering the vehicle as a single rigid body that moves with planar motion, being the wheels are anchored to the frame by rigid bearings, with the possibility of steering only on the front wheels. The following figure schematizes the four slip angles of the tires which, since the z axis is facing upwards, are all negative (clockwise rotation).

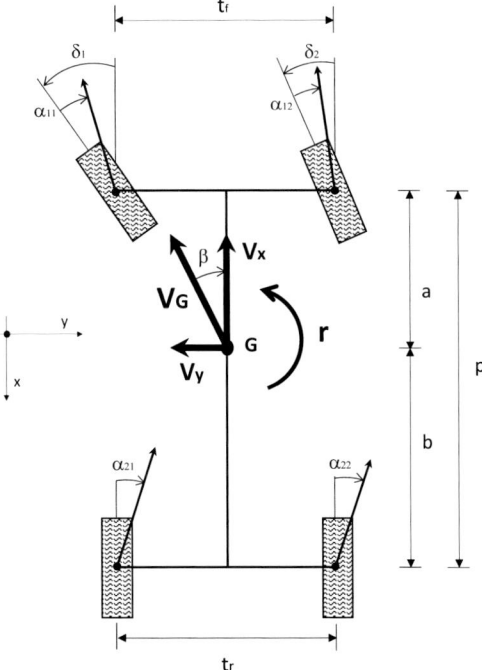

Figure1 – Equivalent quadricycle with the velocity vectors of the wheel centers.

4.2. The evaluation of the slip angles

The estimation of slip angles is an essential step to interpret the vehicle dynamics because they represent the results of the application of the forces on the wheels arising from the vehicle operation.

With reference to the previous figure, the velocity components of the center of gravity G represent the absolute values projected on the reference relative to the vehicle. The resulting formulae are:

$$\mathbf{V}_G \ [m/s] = V_x \mathbf{i} + V_y \mathbf{j} \quad (4.1)$$

$$\Omega = (0, 0, r) \quad (4.2)$$

where $\mathbf{i}, \mathbf{j}$ are respectively the versors of the x axis and the y axis and Ω is the angular velocity of the planar motion.

The presence of a velocity component in y highlights that the vehicle center line does not remain tangent to the trajectory of the center of gravity but it has an angle, called sideslip angle β, which can assume both positive and negative values. Below the formulation of the sideslip angle.

$$\beta \ [radians] = tan^{-1} \frac{V_y}{V_x} \quad (4.3)$$

The control of the sideslip angle is very important for evaluating the dynamic behavior of the vehicle because, as we will see later, it represents a measure of how the rear axle behaves under lateral load.

The study of the equivalent quadricycle, although a more complex schematization than the one we will see later, is of fundamental importance for the evaluation of the slip angles of the four wheels as a function of the kinematic quantities V_x, V_y, r and of the dimensions of the vehicle. The rise of the slip angle on a wheel depends on the forces involved and on the elasticity of the tire, but we can also evaluate it through the aforementioned kinematic quantities which represent the effect of the action of the forces. In fact, the slip angles are the ratios between the transverse and longitudinal speeds of the wheel centers and depend on how the vehicle moves and therefore on V_x, V_y and r. Below the vectors of the velocities of the wheel centers:

$$\mathbf{V}_{11} = (V_x + \frac{rt_f}{2}, V_y - ra)$$

$$\mathbf{V}_{12} = (V_x - \frac{rt_f}{2}, V_y - ra)$$

$$\mathbf{V}_{21} = (V_x + \frac{rt_r}{2}, V_y + rb)$$

$$\mathbf{V}_{22} = (V_x - \frac{rt_r}{2}, V_y + rb)$$

from which the following formulae derive:

$$\tan(\delta_1 - \alpha_{11}) = \frac{V_y - ra}{V_x + rt_f/2}$$

$$\tan(\delta_2 - \alpha_{12}) = \frac{V_y - ra}{V_x - rt_f/2}$$

$$\tan(\alpha_{21}) = \frac{V_y + rb}{V_x + rt_r/2}$$

$$\tan(\alpha_{22}) = \frac{V_y + rb}{V_x - rt_r/2}$$

by making them explicit we obtain equations 4.4

$$\alpha_{11} = \delta_1 - \tan^{-1}\frac{V_y - ra}{V_x + rt_f/2}$$

$$\alpha_{12} = \delta_2 - \tan^{-1}\frac{V_y - ra}{V_x - rt_f/2}$$

$$\alpha_{21} = \tan^{-1}\frac{V_y + rb}{V_x + rt_r/2}$$

$$\alpha_{22} = \tan^{-1}\frac{V_y + rb}{V_x - rt_r/2}$$

The schematization just seen assumes the dependence of slip angles on the roll, pitch and load transfers. The slip angles depend on the load transfers and calculating them by measuring the magnitudes V_x, V_y and r it means taking this dependence into account. If, for example, the roll stiffness of the rear axle is increased, causing a greater load transfer, the rear slip angles increase and affect the r which, in turn, increases because the vehicle yaws faster.

It is important to reiterate that the slip angles thus calculated have no degree of approximation nor simplifying hypothesis and are also valid when the vehicle spins. The equivalent quadricycle is therefore an excellent schematization to evaluate, in a fairly precise manner, the trend of the slip angles and the kinematics of the vehicle regardless of the triggering causes.

We shall use the above schematization to demonstrate that the slip angles of the same axle can be considered almost equal. The following relationship:

$$V_x \gg rt/2$$

which is almost always verified (especially at high speeds or with large radii of curvature), allows us to neglect the quantities related to the yaw rate in the denominators of the equations. Consequently, the slip angles of the same axle are almost equal, and the approximate expressions are as follows:

$$\alpha_f = \delta - \tan^{-1}\frac{V_y - ra}{V_x}$$

$$\alpha_r = \tan^{-1}\frac{V_y + rb}{V_x}$$

where α_f and α_r are respectively the slip angle of the front axle and the slip angle of the rear axle and δ is the average of the steer angles.

4.3. The acceleration and the evaluation of center of gravity velocity

Considering the schematization of figure1, we can evaluate the lateral acceleration A_G of the center of gravity by deriving the vector velocity V_G

$$A_G\ [m/s^2] = \frac{dV_G}{dt} = \frac{d(V_x i + V_y j)}{dt}$$

carrying out the calculations and indicating the cross product with the operator ×:

$$\frac{d(V_x i + V_y j)}{dt} = \frac{dV_x}{dt}i + V_x\frac{di}{dt} + \frac{dV_y}{dt}j + V_y\frac{dj}{dt}$$

$$A_G = \frac{dV_x}{dt}i + V_x \Omega \times i + \frac{dV_y}{dt}j + V_y \Omega \times j$$

$$A_G = \frac{dV_x}{dt}i + V_x r j + \frac{dV_y}{dt}j - V_y r i$$

we obtain the components of the acceleration in the two directions (4.5)

$$A_{Gx} = \frac{dV_x}{dt} - V_y r$$

$$A_{Gy} = V_x r + \frac{dV_y}{dt}$$

being

$$\beta = \frac{V_y}{V_x}$$

$$V_y = \beta V_x$$

$$\frac{dV_y}{dt} = \dot{\beta}V_x + \beta\dot{V}_x$$

we get the frequently used relationship

$$A_{Gy} = V_x r + \dot{\beta}V_x + \beta\dot{V}_x$$

which becomes, at constant velocity:

$$A_{Gy} = V_x r + \dot{\beta}V_x$$

If we have a device capable of measuring the components A_{Gx} and A_{Gy} with the accelerometers, and the yaw speed r with a gyroscope, we can integrate the system of the two differential equations 4.5.

$$A_{Gx} = \frac{dV_x}{dt} - V_y r$$

$$A_{Gy} = V_x r + \frac{dV_y}{dt}$$

with V_x and V_y as unknowns, and calculate the four slip angles and the sideslip angle by using the equations 4.4 and 4.3. We can also calculate, in steady state condition, the radius of curvature of the trajectory by applying the following relationship:

$$R[m] = \frac{\sqrt{V_x^2 + V_y^2}}{r} \qquad (4.6)$$

The radius R is greater than zero for the curves to the left and less than zero for the curves to the right.

This is a good example of an experimental approach in which, starting from accelerometric and gyroscopic signals, it is possible to evaluate the slip angles of the axles and therefore the dynamics of the entire vehicle. It is important to pay attention to the fact that this method of evaluating slip angles shows consistent results only for small stretches of road and, since the accelerometric signals are affected by the inclinations of the vehicle, only on perfectly flat roads.

Automotive suspension

4.4. The single-track model

The numeric integration of the two differential equations just seen allows us to determine the slip angles. The next step is the evaluation of the relationship between the slip angles, the steering wheel angle, the sideslip angle and the curvature radius of the trajectory. To do this we must simplify the model by switching from the equivalent quadricycle to the single-track model or equivalent bicycle described in the following figure. The lines 1,2 and 3, for the hypothesis of plane motion, are respectively perpendicular to the velocity of the front wheel center, the center of gravity and the rear wheel center and identify the center of instantaneous rotation *I.C.*. The displacement of any point of the vehicle is a rotation r around the *I.C.*.

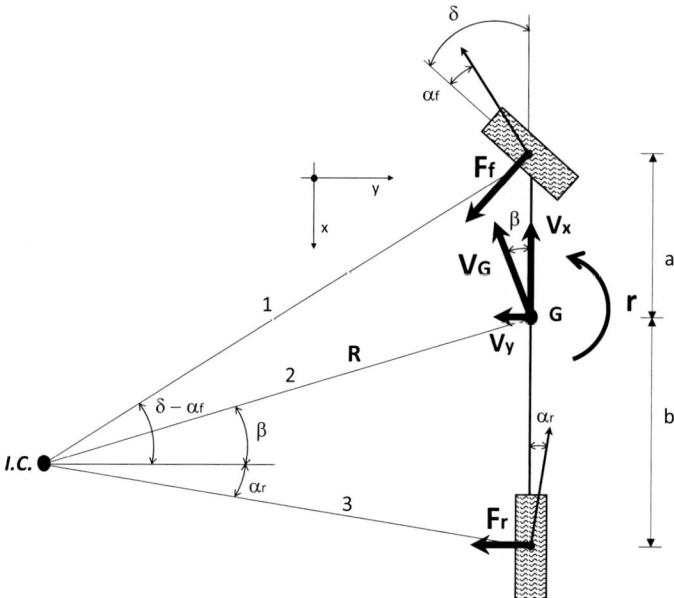

Figure 2 – Single-track model or equivalent bicycle.

This schematization foresees the following hypotheses:
1. The tires of the same axle are condensed into a single tire that has the slip angle equal to the average between the two.
2. The characteristic curve of the single tire of each axle depends on the load transfer and it is built as shown at the end of chapter 2.
3. The steering angle of the wheels is equal to the average between the left and right angles.
4. The angles δ and α are small to approximate, in the various expressions, the cosine to 1 and the sine and the tangent to the corresponding angles measured in radians.
5. The radii of curvature of the trajectories are very high.
6. Tires have a linear elastic behavior.

The hypotheses 4,5 and 6 are fundamental for a quick estimate of the transverse forces but not always for the estimate of the kinematic quantities.

By supposing that the vehicle is travelling a curve at constant speed, with the single-track model we can quickly calculate the centrifugal force distribution on the two axles by following the criteria of the *simply supported beam*, according to which the most stressed axle is the closest to the center of gravity. In this way we get the 4.7

$$F_{y,f} [N] = MA_{Gy} \frac{b}{p} = M_f A_{Gy} \quad (4.7)$$

$$F_{y,r} [N] = MA_{Gy} \frac{a}{p} = M_r A_{Gy} \quad (4.7)$$

with M_f e M_r respectively the front mass and the rear mass.

If the tires have a linear behavior, given the stiffnesses, we can derive the expressions 4.8 of the front and rear slip angles considering that $F_y = C_\alpha \alpha$:

$$\alpha_f [radians] = \frac{M_f A_{Gy}}{C_{\alpha,f}} \quad (4.8)$$

$$\alpha_r [radians] = \frac{M_r A_{Gy}}{C_{\alpha,r}} \quad (4.8)$$

From these expressions it is clear that, with the same value of C_α, the axle closest to the center of gravity is the one that drifts the most, because it is most stressed. One way to bring the angle values closer is to increase the slip stiffness of the most stressed axle by mounting tires with a greater section width. A typical example is represented by some two-seater cars which, having the engine at the rear, have larger sections there.

Obviously, this is true in stationary conditions. In transient conditions the trend of the transverse forces will tend to the steady values undergoing a different progressive trend on each axle. In particular, the axle that first reaches the maximum lateral load is the axle with the greatest roll stiffness.

In the following figure the kinematic single-track model is represented. It has zero slip angles and is very useful for subsequent formulations in which the hypotheses of small angles and large radii hold.

Automotive suspension

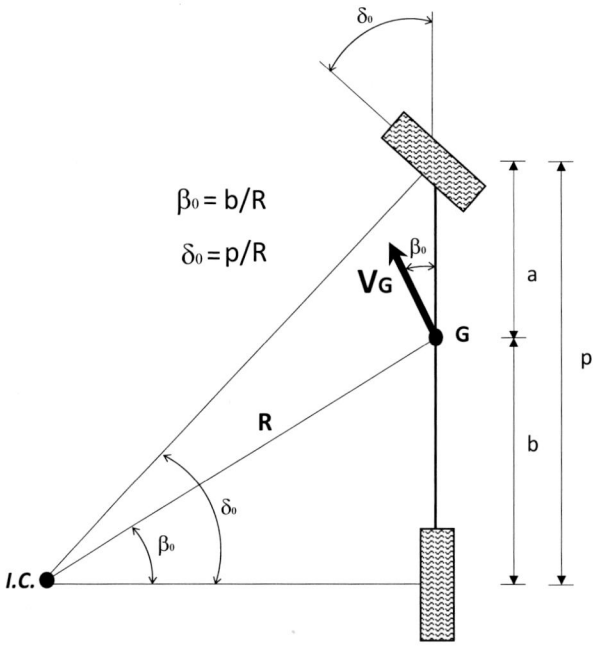

Figure 3 – Kinematic single-track model with sideslip and steering angle values in the hypothesis of small angles and large radii.

4.5. The sideslip angle gradient

With reference to figure 2 we can write the following equation:

$$R \cos \beta \tan \alpha_r = b - R \sin \beta$$

from which:

$$\tan \alpha_r = \frac{b - R \sin \beta}{R \cos \beta} \quad (4.9)$$

The expression 4.9 is true even if the hypotheses 4 and 5 of the previous paragraph (small angles, large radii) are not verified and can represent an excellent kinematic approximation to calculate the sideslip angle knowing the average of the rear slip angles. If indeed β and α_r are very small, we can replace the tangents with the corresponding angles:

$$\beta = \frac{b}{R} - \alpha_r \quad (4.10)$$

in conditions of linear characteristic tires, a condition that is certainly true for small slip angles, it follows that:

$$\alpha_r = A_{Gy} \frac{Mr}{C\alpha, r}$$

and replacing this in the previous equation this results in the following formula:

$$\beta = \beta 0 - A_{Gy} KBETA \quad (4.11)$$

in which $\beta 0$ represents the kinematic value of the sideslip angle, that is the value obtained when the slip angles are zero as shown in figure 3.

KBETA is the sideslip gradient which is inversely proportional to the slip stiffness of the rear axle and directly proportional to the percentage of mass present on the rear axle

$$KBETA = \frac{Mr}{C\alpha, r}$$

From expression 4.11 it is clear that, while the lateral acceleration increases, the sideslip angle starts from the kinematic value $\beta 0$, decreases, becomes zero and then changes sign. The following figure shows the trend of β in a Steering pad maneuver, that is, a curve with a constant radius with a progressive increase in speed.

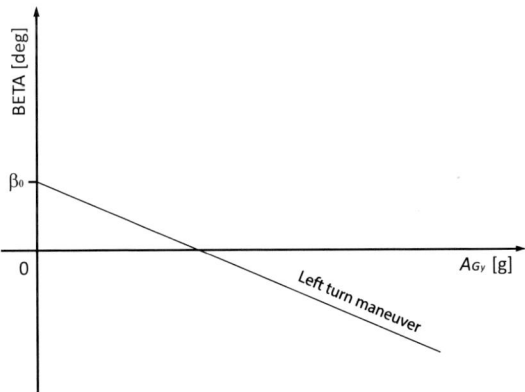

Figure 4 – Sideslip angle trend in a steering pad maneuver in the hypothesis of linear tires.

Below is the speed at which it is canceled:

$$V, \beta nullo = \sqrt{\frac{bC\alpha, r}{Mr}}$$

which is obtained by writing R as a function of the speed in the expression of $\beta 0$ of 4.11 as follows:

$$R = \frac{V^2}{A_{Gy}}$$

Automotive suspension

The linear trend shown in figure 4 is true only under the hypothesis of linear characteristics of the tires (which is generally true for small slip angles). Otherwise, the shape remains linear only for small accelerations which always correspond to small slip angles. Below is the graphical representation of the sideslip angle complete with sign, for speeds lower and higher than $V,\beta null$.

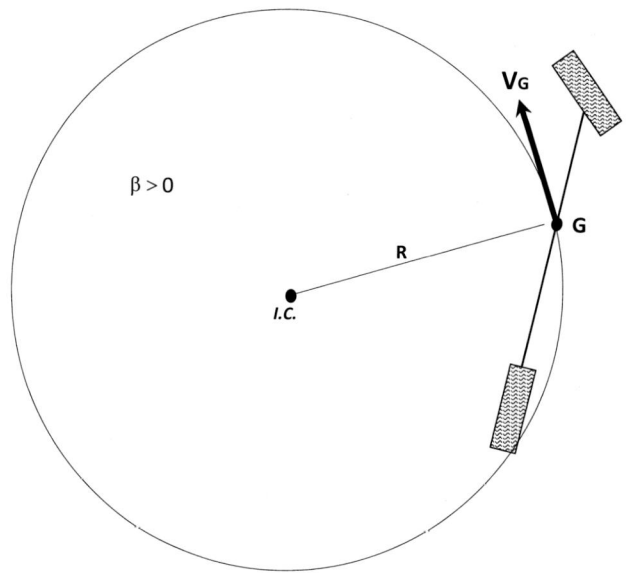

Figure 5 – Sideslip angle for speed less than $V,\beta null$ in stationary conditions.

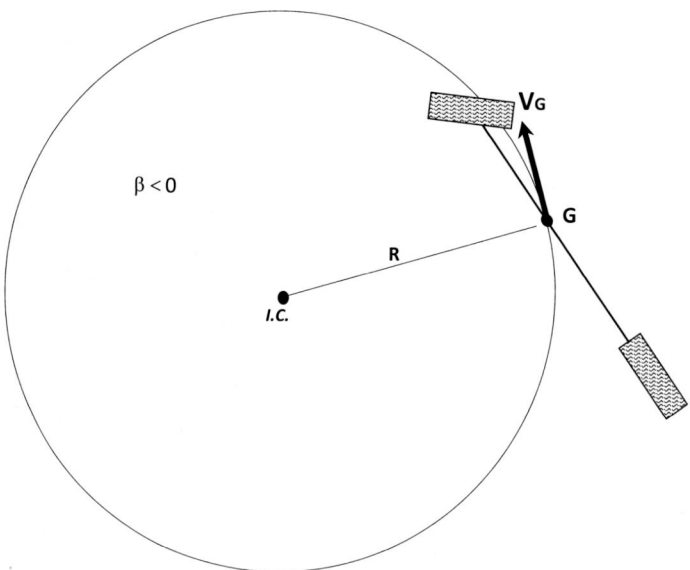

Figure 6 – Sideslip angle for speed greater than $V,\beta null$ in stationary conditions.

4.6. The understeer gradient

Always in the hypothesis of a steady state condition, the single-track model allows us to write the following relationship between the slip angles, the steer angle and the radius of curvature.

$$\tan(\delta - \alpha_f) = \frac{a + R \sin \beta}{R \cos \beta}$$

This expression is true even if the hypotheses 4 and 5 of the single-track model are not verified (small angles, large radii) and represents an excellent kinematic approximation in order to calculate the sideslip angle, knowing the average of the front slip angles and the average of the steering angles. Under the hypothesis of small angles and large radii of curvature and considering 4.10, the formula becomes:

$$\delta = \frac{p}{R} + (\alpha_f - \alpha_r) \quad (4.12)$$

which on the steering wheel angle (*SWA*) results:

$$SWA = \frac{p}{R}\tau + (\alpha_f - \alpha_r)\tau$$

Where the *SWA* is the steering wheel angle and τ is the steering ratio equal to the kinematic ratio between the steering wheel angle and the average of the wheel angles (δ). In general, the steering ratio τ is around 15-16 for mass-produced cars and around 12-13 for sports cars.

By replacing the 4.8 in the previous equation we get:

$$SWA = \frac{p}{R}\tau + A_{Gy}\left(\frac{M_f}{C_{\alpha,f}} - \frac{M_r}{C_{\alpha,r}}\right)\tau$$

and synthesizing we obtain the equation:

$$SWA = \frac{p}{R}\tau + A_{Gy}Kus \quad (4.13)$$

Where *Kus* is the understeer gradient which represents the increment of the steering wheel angle necessary to have a lateral acceleration increase equal to 1 m/s² in a constant radius curve.

$$Kus\ [rad/(m/s^2)] = \left(\frac{M_f}{C_{\alpha,f}} - \frac{M_r}{C_{\alpha,r}}\right)\tau$$

Referring again to a constant radius curve in stationary conditions, based on the value of this quantity the vehicle will be:

<u>Understeer if Kus>0:</u>

The vehicle is understeer when an increase in lateral acceleration corresponds to an increase in the steering wheel angle. In a curve with a constant radius, a delta A_{Gy}, due to an increase in speed, produces an increase in the steering wheel angle caused by the increase of the second member of 4.13. This is because the growth of the front slip angle is greater than that of the rear slip angle. Understeer behavior is a condition of stability. It is what is required from the vehicle since it is the behavior closest to the user instinct which, as speed increases, tends to steer more.

Oversteer if Kus<0:

The vehicle is oversteer when an increase in lateral acceleration corresponds to a decrease in the steering wheel angle. The oversteer behavior is a condition of instability. This because the need to lower the steering wheel angle depends on the fact that the increase in lateral acceleration, due to an increase in speed, causes a decrease in the radius of curvature and consequently a further increase in the lateral acceleration, following the law

$$A_{Gy} = \frac{V^2}{R}$$

The user must counter-steer to cancel this effect. From an analytical point of view $K_{us} < 0$ means $\alpha_r > \alpha_f$ and when the rear slip angle is greater than the front slip angle, the necessary steering angle is less than the kinematic one, which is p/R. By starting from the kinematic steering condition, an increase in acceleration corresponds to a decrease in the initial steering wheel angle.

Neutral if Kus=0

It is a behavior that is difficult to achieve because whatever the lateral acceleration is, it must always be verified that $\alpha_f = \alpha_r$.

Below is the trend of the steering wheel angle during a curve maneuver with a constant radius at slow increasing speed, under the three conditions described:

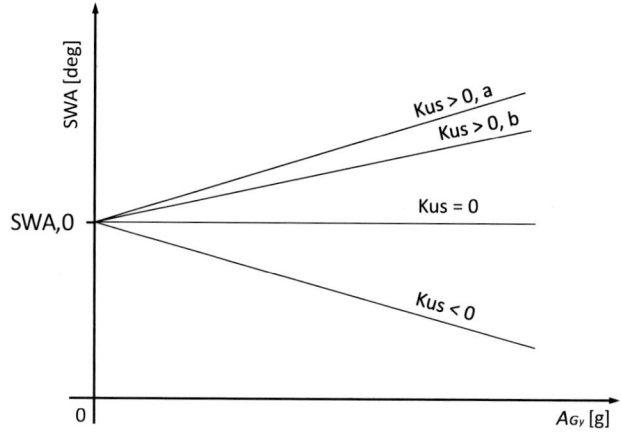

Figure 7 – Understeer curves with different slopes.

The curves are called understeer curves. *Kus* represents the slope and *SWA*, 0 represents the kinematic steering wheel angle shown in figure 3. The two curves *a* and *b*, both with positive slope, represent ordinary conditions. The other two curves at *Kus* = 0 and *Kus* <0 represent conditions from which to keep away. The difference between the two curves *a* and *b* lies in the different response to the increase in speed or in the sensitivity of the steering to increasing lateral acceleration. In particular, curve *b* is the understeer curve of a more direct vehicle because it requires a smaller steering wheel angle for the same lateral acceleration and radius of curvature. To go from curve *a* to curve *b* it is required to work on the suspension and steering, as we will see in chapter 6.

From an experimental measurement of the slip angles using equations 4.4, we can evaluate the oversteer or understeer of the vehicle by calculating the *Kus* starting from $(\alpha_f - \alpha_r)$ being:

$$\alpha_f = \frac{(\alpha_{11} + \alpha_{12})}{2} \qquad \alpha_r = \frac{(\alpha_{21} + \alpha_{22})}{2}$$

keeping in mind the (4.8)

$$(\alpha_f - \alpha_r) = A_{Gy} \left(\frac{M_f}{C_{\alpha,f}} - \frac{M_r}{C_{\alpha,r}} \right)$$

multiplying the two members by the steering ratio, we get:

$$(\alpha_f - \alpha_r)\, \tau = A_{Gy} Kus$$

$$Kus\ [rad/(m/s^2)] = \tau \frac{(\alpha_f - \alpha_r)}{A_{Gy}} \qquad (4.14)$$

in deg/g:

$$Kus\ [deg/g] = \tau \frac{deg\,(\alpha_f - \alpha_r)}{A_{Gy}/9.81} \qquad (4.14)$$

According to what has been said, it is possible to calculate the understeer gradient by experimentally measuring the slip angles or by integrating equations 4.5 and then by inserting the results into 4.4. For the calculated value to be correct, the car must be stabilized in cornering at constant speed, constant roll and a maximum lateral acceleration of 0.3-0.4 g, so that the tires still have linear behavior and that the angles involved are small. An effective test to evaluate the correctness of the calculated *Kus* is to compare the value obtained with the understeer gradient at constant speed, which can be evaluated in the final section of a step steer (see chapter 8), by computing the relationship between the steering wheel angle and lateral acceleration, as follows:

$$Kus,cv \text{ [deg/g]} = \frac{SWA}{A_{Gy}} 9.81$$

where SWA and A_{Gy} are always measured with stabilized vehicle and the lateral acceleration does not exceed 0.4g. The understeer gradient at constant speed is calculated starting from straight-line driving. It does not depend on speed application of the steering wheel angle, with which the step steer is taken, and it is always greater than the constant radius understeer gradient. Regardless of the demonstration, it turns out to be:

$$Kus,cv = Kus + \tau \frac{p}{V^2} \frac{180}{\pi} 9.81$$

if the equality exists, the evaluation of Kus is correct and consequently also that of the slip angles.

In non-stabilized conditions and with accelerations greater than 0.3-0.4 g (non-linearity of the tires) we can only calculate the difference in the slip angles of the two axles to understand if the attitude of the vehicle, in certain points of the circuit, is over or understeer and intervene accordingly but we are far from the evaluation of the understeer gradient.

4.7. No stationary conditions

In a generic curve with constant radius and variable speed, the presence of an angular acceleration $\dot{r}$ produces a yaw moment that we can approximate as:

$$M,yaw \text{ [Nm]} = J\dot{r}$$

In which J is the inertial moment of the vehicle around the z axis. The approximation lies in the fact that, in order not to introduce excessive complications that would not significantly change the final result, the effect of the inertial tensor on M,yaw is not considered. This effect is exhibited with yawing gyroscopic torques, generated during the roll and pitch of the car, which add up to the aforementioned torque. With the simplification made, the expression of the transverse force on each axle that balances the yaw moment is the following:

$$Fyi \text{[N]} = \frac{J\dot{r}}{p}$$

In case of an increment of speed, since the angular acceleration increases, the yaw moment has the direction concordant to the direction of the curve, the emerging force on the front axle is oriented towards the center of curvature and the force on rear axle has opposite sign. The formulae are thus:

$$Fyi,f \text{[N]} = \frac{J\dot{r}}{p}$$

$$F_{yi,r}[N] = -\frac{J\dot{r}}{p}$$

To analyze in detail the cornering behavior at constant speed, which is always characterized by an increase and a subsequent decrease in angular acceleration, we shall refer to the step steer. In the following figure, the typical trend of yaw speed in a step steer at constant speed with two different velocities of application of the steering wheel angle is shown. The red curve corresponds to the application of a steering wheel angle of 20 degrees according to a linear ramp lasting 0.5 seconds while the one in green according to a linear ramp lasting 0.2 seconds.

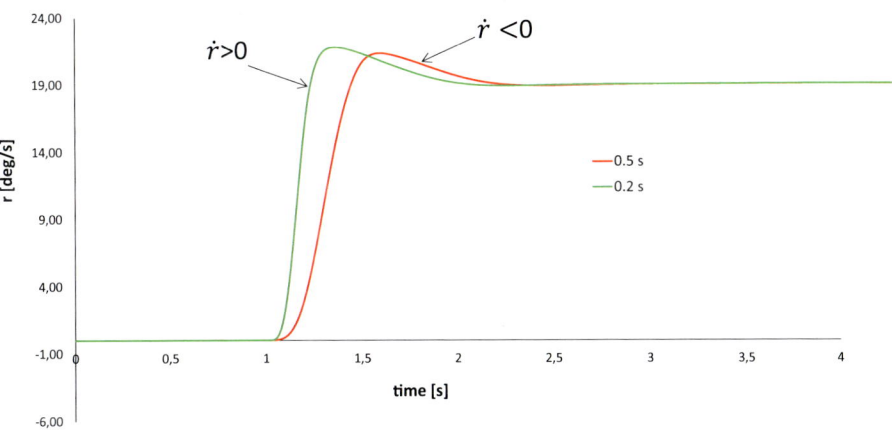

Figure 8 – Trend of yaw speed in a step steer maneuver at 100km/h with the application of two 0.2s and 0.5s linear ramps. The step steer maneuver is an excellent way to understand the problems of the first corner entry.

This maneuver, which consists in the sudden application of a certain steering wheel angle to a car launched with a certain speed, is the one that best approximates the corner entry (see chapter 1). From the figure it is very clear that, in the initial phase, the yaw rate r is increasing, while in the second phase it is decreasing. The first phase corresponds, for the above reasons, to a request for an increase in force on the front axle, which increases as the speed of application of the steering wheel angle increases. The production of this increase is entrusted to the front axle which, if it does not have a correctly sized suspension, does not allow the insertion speed desired by the driver, causing the car to understeer.

A correctly sized suspension is able to orient the wheels to produce a transverse force, consistent with the required angular acceleration aligned to the vehicle mission, with a slip angle (imposed by the steering wheel on the tires) less than the *sliding limit* of the characteristic emerging axle (see chapter 2). In the second part of the step steer, as in cornering, the yaw moment becomes negative and the rear responds with an increase in lateral force directed towards the center of the curve. The car goes into spin if this last increase is not allowed by the rear axle architecture. Contrarily the entry into the curve is fast if the achievement of the required increase in force is sudden, as will be clearer in chapter 8. In the following figure, the schematic diagram of the temporal arrangement of the intervention of the lateral forces on the two axles is shown. The graph shows approximately the time intervals relating to the intervention of the lateral forces; they overlap to the extent that the rear is quick to react to the force that arises on the front. The first part of the yaw speed, with positive angular acceleration, corresponds to the request for a lateral force (in transparent red) on the front axle whose peak is greater than the lateral force (in red) related to the stabilized condition with constant radius and zero angular acceleration. The second part of the yaw speed, with negative angular acceleration, corresponds to the request for a lateral force (in transparent blue) on the rear axle whose peak is greater than the lateral force (in blue) related to the stabilized condition.

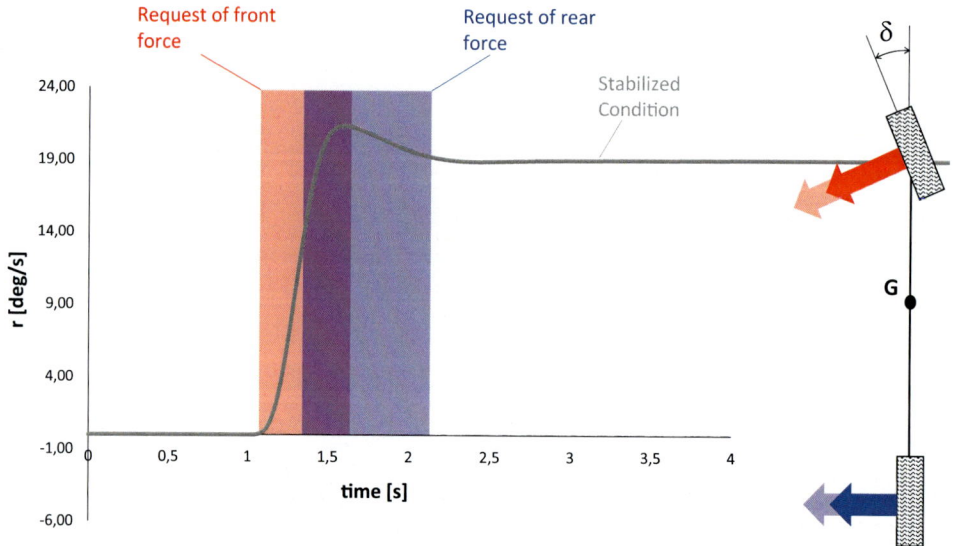

Figure 9 – Approximate temporal arrangement of the lateral force request on the two axles.

The general behavior of the vehicle at constant speed can be schematized with good approximation by superimposing the effect of the centrifugal force and the effect of the angular acceleration, as shown in the following figure.

Lateral dynamics – Use of analytical models

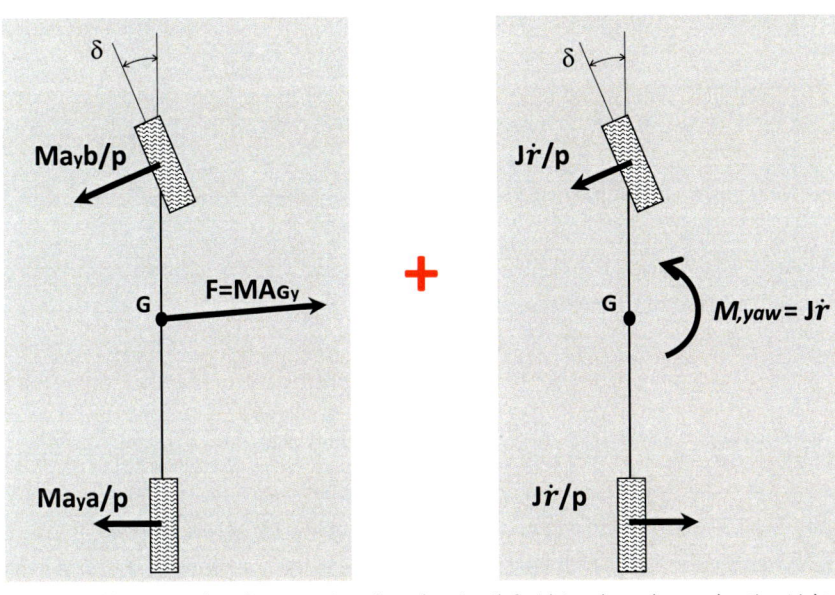

Figure 10 – Lateral forces resulting from centripetal acceleration (left side) and angular acceleration (right side). In the latter case *M,yaw* is the resultant Moment acting on the vehicle and not the Moment of the forces of inertia.

By adding together the forces deriving from the centripetal acceleration in a stationary conditions and the forces deriving from the angular acceleration (taking into account the hypotheses 4,5 and 6 of paragraph 4.4) we obtain the expressions of the forces on each axle:

$$Fy,f[N] = MA_{Gy}\frac{b}{p} + \frac{J\dot{r}}{p}$$

$$Fy,r[N] = MA_{Gy}\frac{a}{p} - \frac{J\dot{r}}{p}$$

from which the corresponding slip angles are obtained:

$$\alpha_f = \frac{Fy,f}{C\alpha,f} = \frac{1}{C\alpha,f}(MA_{Gy}\frac{b}{p} + \frac{J\dot{r}}{p})$$

$$\alpha_r = \frac{Fy,r}{C\alpha,r} = \frac{1}{C\alpha,r}(MA_{Gy}\frac{a}{p} - \frac{J\dot{r}}{p})$$

Now we can derive the steering wheel angle, as a function of the slip angles, by applying the following formulation

$$SWA = \frac{p}{R}\tau + (\alpha_f - \alpha_r)\tau$$

By making explicit the slip angles we obtain

$$SWA = \frac{p}{R}\tau + A_{Gy}\left(\frac{M_f}{C_{\alpha,f}} - \frac{M_r}{C_{\alpha,r}}\right)\tau + \left(\frac{1}{C_{\alpha,f}} + \frac{1}{C_{\alpha,r}}\right)\frac{J\dot{r}}{p}\tau$$

Finally, by replacing the K_{us} we obtain the equation (4.15)

$$SWA = \frac{p}{R}\tau + A_{Gy}\,K_{us} + \left(\frac{1}{C_{\alpha,f}} + \frac{1}{C_{\alpha,r}}\right)\frac{J\dot{r}}{p}\tau$$

which is equivalent to (4.13) with the addition of the non-stationary term. The use of this equation, which has a physical meaning when the hypotheses for the superimposition of the effects are considered, presupposes the knowledge of the temporal trend of the lateral acceleration, of the yaw acceleration, of the moment of inertia J, of the slip stiffnesses etc.. For this reason, in order to justify some phenomena, we will only give a qualitative interpretation of its structure. However, the reader can engage in a quantitative verification in a *multibody* computing environment. In this environment, which we will describe in more detail in chapter 8, by simulating a step steer maneuver, the accuracy of equality 4.15 can be verified by comparing the trend over time of the first member, which is known because it is imposed by the user, with the quantities calculated on the second member, which derive from the measurements made on the virtual model. Numerical calculation is a great help to validate mathematical formulations that otherwise would have to be managed analytically with many difficulties. We explain equation 4.15 below.

The non-stationary term of equation (4.15) quantifies not only the vehicle agility in corner entry but also the lateral force surplus initially required on the front axle and subsequently on the rear axle. Cars with a weight distribution shifted towards the rear, which generally suffer from oversteer in a stabilized mode, as they are unable to react to the excess force required from the front axle due to the contained vertical load on the tires, risk understeer in corner entry. The request for lateral force on the two axles, with the same angular acceleration, is reduced by the reduction of J and does not depend on the slip stiffness. The steering wheel angle necessary for a certain positive angular acceleration, corresponding to the first corner entry or to an increase in speed when cornering, is reduced when the moment of inertia J is lowered. It is important to pay attention to the fact that an increment of wheelbase corresponds to a reduction of the non-stationary term only if it does not generate a significant increment in J.

Continuing the analysis of the non-stationary term of equation (4.15), it can be noted that the steering wheel angle request is inversely proportional to the sum of the slip stiffnesses and does not depend on their distribution, as happens in the stationary case. This behavior, predictable for the front axle, results from the fact that, in the presence of angular acceleration, the forces required on the axles have the opposite sign. In particular, in the case of positive angular acceleration, the force that emerges on the front axle is directed towards the center of the curve while that which emerges on the rear axle is directed towards the outside of the curve. Analyzing the case of angular acceleration in a constant radius curve, which is certainly more intuitive, we can say that the front axle produces an increase in the slip angle while the rear axle produces a decrease in the slip angle, as shown in the following figure. Therefore, both axles contribute to an increase in the steering wheel angle. If, for example, the slip stiffness of the rear axle increases, the corresponding decrease in slip is lowered and the vehicle requires a lower steering wheel angle delta than in the case of a less rigid axle. It shall be reiterated that this effect, which takes into account only the angular acceleration, is superimposed on the effect of the lateral acceleration which always produces slip increases on the two axles.

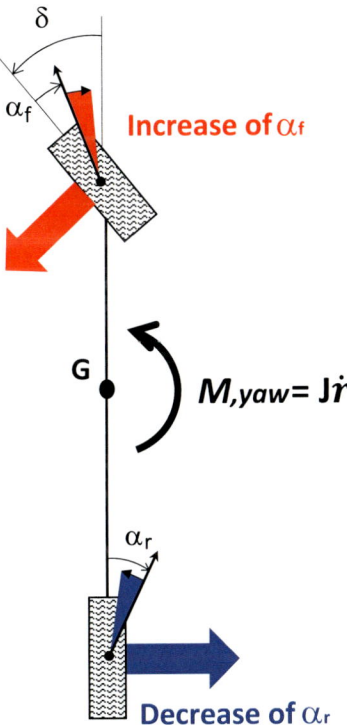

Figure 11 – Lateral forces resulting from angular acceleration with the effect on the corresponding slip angles.

4.8. Summary

The analytical formulations of the second paragraph allow us to evaluate the slip angles of the four wheels starting from the knowledge of V_x and V_y of the center of gravity and of the yaw rate r. The two components of the speed according to x and y can be obtained (for small sections and flat roads) from the integration of equations 4.5 which assume the measurement of the longitudinal and lateral acceleration of the center of gravity, the yaw speed and the steering wheel angle. The calculation of slip angles allows us to make important assessments on the dynamic behavior of a vehicle.

At constant speed, the lateral acceleration consists of a first contribution which depends on the yaw speed and a second contribution which depends on the variation of the sideslip angle and is canceled if the latter stabilizes.

The single-track model allows the interpretation of the understeer and oversteer of the vehicle based on the difference between the front axle slip angle and the rear axle slip angle. This last difference, responsible for the performance of the understeer curve, depends on the operating range of the front and rear tires at different lateral acceleration. In stationary conditions, the sideslip angle is only a function of the rear axle slip.

In a step steer, the lateral force request involves first the front axle and then the rear axle. The two axles react to these forces, whose peaks are greater than the stationary values, thanks to the architecture and the tuning of the suspensions. If the two axles are unable to exert the required forces, understeer occurs if the lack is found at the front, while oversteer occurs if the lack is found at the rear. The surplus of lateral force required from the two axles is lowered when the moment of inertia J around the z axis is reduced while the increase in the steering wheel angle, generally required in transients, decreases as the sum of the slip stiffnesses of the two axles increases.

5. SUSPENSION
Architectures and schematizations

The first distinction, relative to the various architectures, concerns independent wheel suspension and dependent wheel suspension. In independent wheel suspension, the movement of each axle wheel does not affect the movement of the other wheel, while in dependent wheel suspension, due to the interconnections between the two wheels, the opposite happens.

The other characteristics that distinguish the different architectures regard above all the kinematics that drives the movement of the wheel with respect to the vehicle body in the different operating conditions. The kinematics is the set of components (arms and rods) and constraints that allow the wheel to assume the desired positions during the operation of the car. In particular, the more the number of components increases, the more the possibility to drive the wheel towards the positions which improve the car performances increase. The study of the relative movement of the wheel with respect to the vehicle body is called elastokinematics and includes both rigid displacements (kinematics) and elastic displacements under various load conditions.

The choice of the architecture depends on the vehicle mission and on the cost. Sports cars prefer architectures that can guarantee elastokinematics capable of increasing dynamic performances with a limited mass. Road cars prefer architectures that guarantee handling performance and *comfort* in line with the category, low production costs and limited dimensions. The latter feature is of primary importance when, in choosing of the rear suspension, the aim is to increase the volume of the car trunk.

5.1. The schematic of the suspensions

To understand the functioning of the suspensions it is necessary to look at the way in which the elastokinematic scheme, consisting of rigid bodies and connection constraints, defines the movement of the wheel. The objective of the schematization of a suspension is the creation of a mathematical model that, according to the constraints and the bodies involved, returns the displacements and the orientations of the various components of the kinematics. We will not go into the details of the motivations but we will only make considerations on how to simplify the components and their joints in order to define the possibilities of relative and absolute movement of the various parts of the suspension. The next step in identifying the information necessary to evaluate the motion of the bodies involved is the construction of a three-dimensional model that can reproduce the dynamic behavior of the suspension.

The elastokinematic schematization of each component is carried out according to well-defined rules. For example, the control arm is schematized or simplified as a rigid body (or as a flexible body in more detailed analyses) represented by a triangle with three constraints at its ends. The constraint is graphically a sphere if the joint of the arm at that point is a ball joint or a cylinder if the joint is an elastic bushing. The dimensions of the triangle and the type of constraints define the elastokinematics of the arm. The following figure shows the drawings of two different schematics of the control arm overlapped with the pictures of the corresponding components. The upper control arm is connected to the vehicle body with two ball joints while the lower control arm with two bushings. Both arms are connected to the upright with ball joints. The connection to the vehicle body with ball joints is used in racing cars while the connection with elastic bushings is used in road cars.

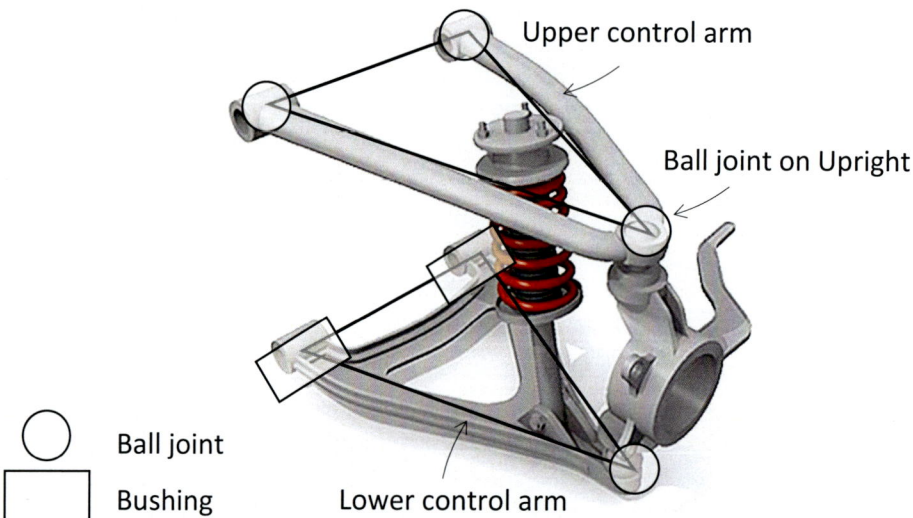

Figure 1 – Example of elastokinematic schematization of the control arms on a low Double wishbone. The upper control arm has been schematized with three ball joints, the lower arm has been schematized with two bushings to the vehicle body and a ball joint on upright

The hub carrier or upright is schematized by a cylinder rigidly connected to three segments. The cylinder represents the interface towards the wheel bearing while the three segments represent the connections to the arms and the steering rod, as shown in the following figure.

Suspension – Architectures and schematizations

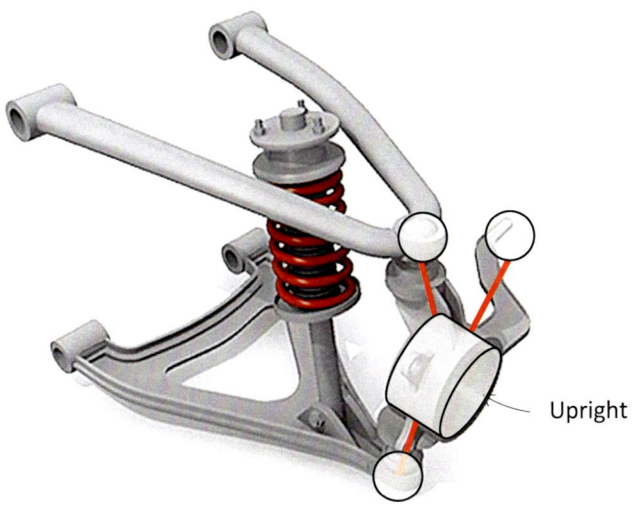

Figure 2 – Example of kinematic schematization of the upright or upright modeled with a cylindrical rigid body parallel to the wheel axis and joined with three ball joints to the arms and to the steering rod.

The connection constraints to the arms and to the steering rod are ball joints in all types of cars. The rim of the wheel is connected to the upright with a hinge (not shown in the figure) which constrains only the rotation around the wheel axis. This hinge represents the wheel bearing, screwed or mounted with interference on the upright. This is schematized with an elastic bushing to simulate its flexibility. The elastic behavior of the wheel bearing must be well designed because it represents the main cause of the loss of *camber* under lateral load, a very important performance that will be fully described in the following chapters.

The generic rods, such as the steering rod and the rods of the multi-link, are always schematized with straight segments identified by the centers of the connecting joints, as shown in the following figure. Also, in this case, the constraints can be ball joints or elastic bushings. Any curvature shape has no kinematic influence but only serves to avoid the interferences with the other elements during the operation of the suspension. In this way, the rod behaves like a simple pendulum that constrains the upright to the vehicle body.

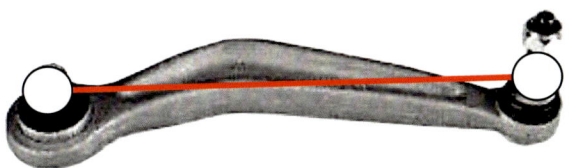

Figure 3 – Kinematic schematization of a rod.

The positions of the joint centers and bushing centers are the main variables of the kinematic because they define the movement of the wheel with respect to the vehicle body. If, for example, the positions of the center joints of a control arm change, its axis of rotation changes and consequently the movement of the whole part and of the kinematic mechanism that contains it.

At the beginning of the paragraph, we stated that the elastokinematic scheme of a suspension consists of rigid bodies and of the constraints connecting them. In reality, the various components, which are generally made of metal (steel, aluminum, cast iron, etc.) or of composite materials, have a certain elasticity that must be considered in very detailed analyses. However, in this text the author has not deemed necessary to consider the flexibility of the components involved because it would lead to unnecessary complications without significantly changing the final results.

5.1.1. The elastic bushings

Elastic bushings are deformable joints that filter the external forces deriving from the operating conditions, to make the vehicle more comfortable. They consist of two sleeves (generally metallic) integral with a rubber element which, according to its composition and size, defines the elastic characteristics of the bushing. The outer sleeve is made integral with the mobile part (arm or rod) and the inner sleeve is made integral with the fixed part (subframe or vehicle body). In the following figure is shown an example of a bushing which allows the rotation around z, usually used to constrain the arms and rods.

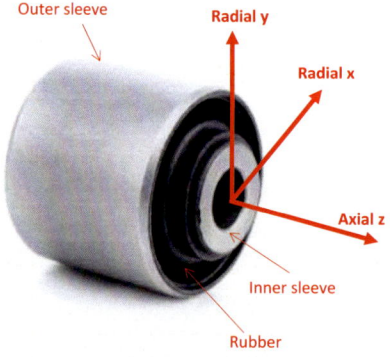

Figure 4 – Elastic bushing with the description of the various components, of the axial direction and of the two radials.

The behavior of the bushings depends on the elastic characteristics in the six directions, that is, on the curves that define the reaction forces and moments as a function of the displacement and relative rotation between the two sleeves. The sizing of the stiffnesses of the bushings is quite complex. In fact, it affects both the method of transmission of forces on the vehicle and the elastic displacements of the suspension components that define the inclinations of the wheel with respect to the vehicle body, tendencies that are generally in contrast with each other. We shall show a very qualitative example to

Suspension – Architectures and schematizations

motivate what has just been said. Using a soft bushing for the rear attachment (22 in the following figure) of the lower control arm, the forces deriving from the road surface would be transmitted more smoothly, producing less noise and vibrations in that region. The same bushing could have an undesirable impact on the elastokinematics due to the excessive wheel opening under lateral load, due to the low stiffness on the point 22. This is because the lateral displacement Sy of point 22 of the arm is inversely proportional to the stiffness Ry of the bushing along the y direction, according to the ratio:

$$Sy = \frac{F22}{Ry}$$

Therefore, the arm rotates around point 11 which moves sideways by a very small amount because its bushing is very rigid in y. This schematization requires the further hypothesis that the radial stiffnesses (y of the car) are much larger than the axial stiffnesses (x of the car) and the conical stiffnesses (moment on relative rotation between the sleeves).

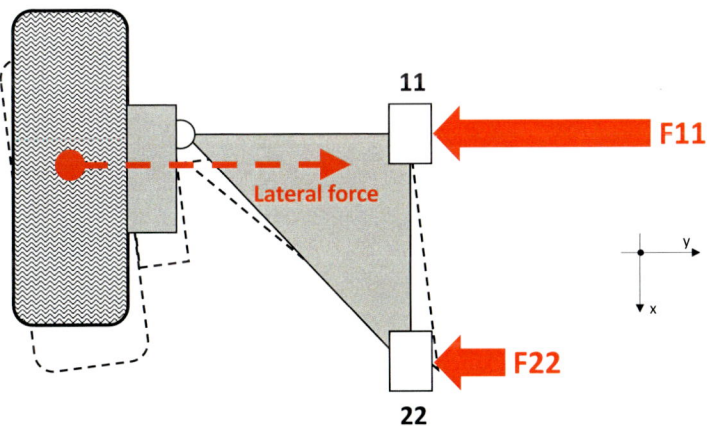

Figure 5 – Left front suspension. Plan view of the lateral forces acting on the arm and deformation of the entire system (dashed).

The situation just shown is easy to analyze because, being reactively isostatic in the transverse direction, it allows to calculate the reaction loads F11 and F22 regardless of the radial stiffness of the two bushings. In this case the value of the force F22, and the consequent displacement Sy, depend only on the distance of the point 22 from the transverse force and not on the stiffness of the other bushing 11. If this distance decreases, the values of F22 and Sy increase (F11 decreases) and consequently the wheel opening increases. These evaluations are certainly more complicated, in a hyperstatic condition because the value of the forces, and of the consequent associated displacements, do not depend only on the geometry but also on the combination of the rigidity of the bushings. We shall imagine, for example, inserting the steering rod as well, simulating the compliance of the rack housing with the bushing 33. If the latter is very soft (compared to the others) the arm rotates as in the previous case, if instead it is very rigid it absorbs almost all the transverse force and, since the displacement of point *22* is lowered, the rotation of the arm is attenuated. In the following figures the comparison between the two deformed ones is shown. It shall be repeated that the examples seen represent only qualitative support for the interpretation of the elastic behavior of a suspension which in its entirety is much more complex and requires the use of dedicated calculation codes.

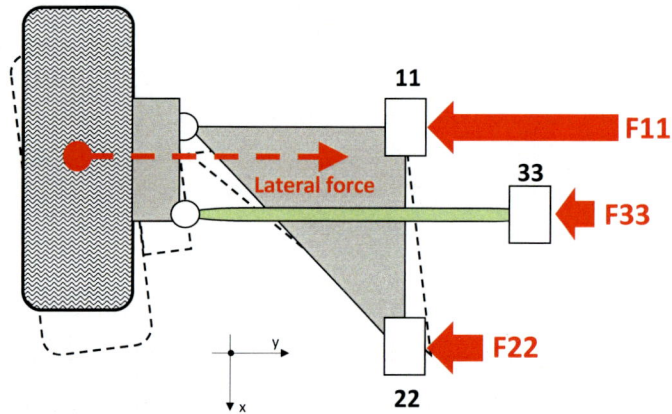

Figure 6 – System deformation in case of soft bush 33.

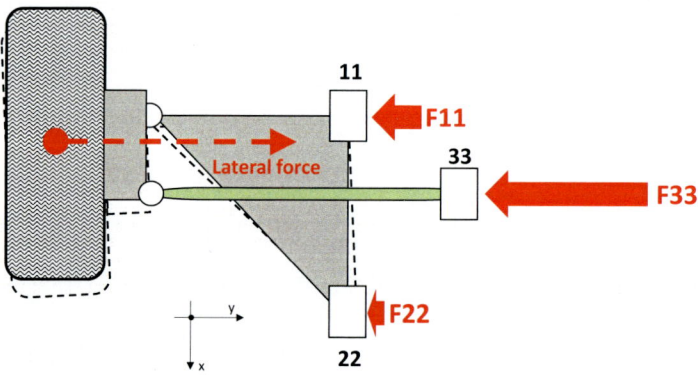

Figure 7 – Deformed system in case of bushing 33 stiffer than 11.

The torsional stiffness of the elastic bushings contributes to the ground stiffness of the suspension (*Ksusp*) seen in chapter 3. Therefore, if elastic bushings are present, it is necessary to consider their torsional stiffness in the calculation of the *wheel rate* or *wheel ride rate*. The elastic bushings are normally absent on sports cars, but they can be used to induce some behaviors such as the closing of the rear wheel under lateral load which accelerates the growth of the transverse force on the rear axle and makes the vehicle faster in the transitory (see chapter 4). The impact of the latter on vehicle dynamics will be fully described in the following chapters.

5.2. Front suspensions

There are different front suspension architectures, both with independent and dependent wheels, chosen according to the vehicle mission. Generally, the front wheel architectures with dependent wheels, thanks to their robustness characteristics, are used for vehicles designed for extreme off-road. Otherwise, for all other uses, independent wheel architectures are preferred. In this text we will describe only the *Double wishbone* suspension and the *McPherson* suspension, normally adopted for most of the cars.

5.2.1. Double wishbone

The Double wishbone suspension is quite complex and expensive because it is equipped with two control arms (lower and upper) that orient the upright in its bump and rebound travel (upward and downward displacement of the wheel). During the bump-rebound travel, the Double wishbone is able to guarantee a good control of the wheel angles variation as *camber* gain and *toe* variation which contribute, as we will see later, to improve the dynamics performance of the vehicle. The presence of two arms gives a better response to the transverse loads such as the reduced loss of *camber* in the cornering. In the following figure the overall view (a) and the exploded view (b) of the schematization of the kinematic mechanism of a Double wishbone are shown.

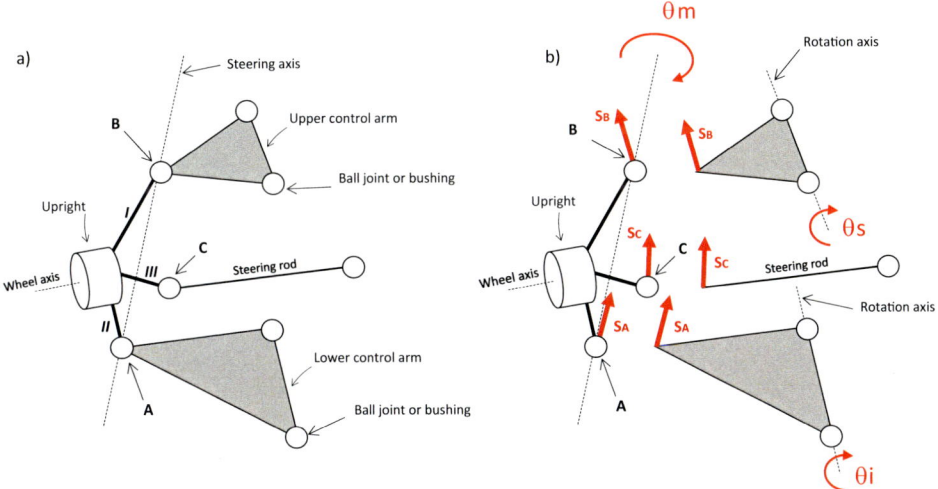

Figure 8 – a) Elastokinematic schematization of a Double wishbone. b) Displacements imposed by the constraints and equality between the displacements of the points common to the contiguous parts.

The arms are connected to the vehicle body either through two ball joints, in the racing cars, or through two elastic bushing, in the road cars. The rotation axis of an arm is the joining between the centers of the corresponding ball joints or the common axis of the two bushings. If the bushings are built with a very rigid material, they must have the rotation axes coincident, otherwise the arm does not move. The upright consists of a cylindrical body (bearing housing) rigidly connected to the three links I, II and III connected to the arms through the ball joint A and B and to the steering rod through the ball joint C. The joining of the joint centers A and B identifies the rotation axis of the upright and therefore the steering axis. In the right side of the previous figure the displacements of the ends of the arms due to the corresponding rotations around their axes are highlighted. In particular, the rotation θi of the lower control arm produces the displacement SA of the extreme of the arm which is equal to the displacement SA of the corresponding point that belongs to the upright. The displacements of the points of the upright are those imposed by the arms, by the steering rod (SA, SB, SC) and those due to the rotation θm around the steering axis.

The Double wishbone suspension guarantees a good control of the location of the steering axis because has good margins to move the joint centers A and B. This possibility is very important because it allows to reduce the transverse inclination of the steering axis (*King Pin Inclination*) which, as we will see in chapter 7, greatly affects the dynamic performance of the vehicle. In particular, in the case of the high Double wishbone, the position of the upper joint center can go outwards without interfering with the rim. In fact, as can be seen from the following figure, the shape of the upright brings the joint center of the upper arm to a greater height than the upper edge of the tire in moving it outwards by the desired amount.

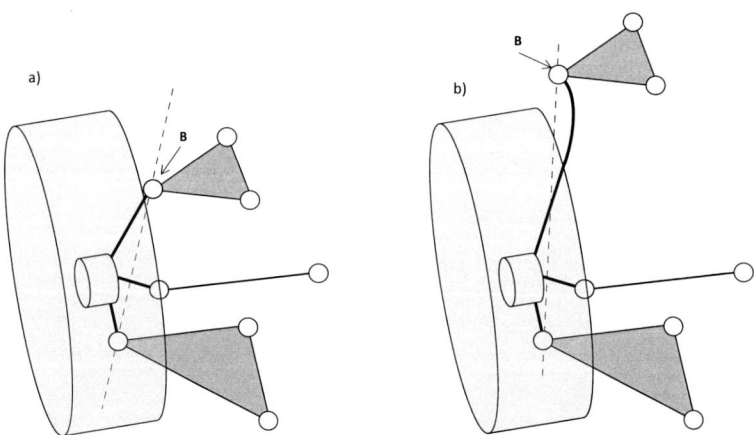

Figure 9 – a) Schematic of low Double wishbone. b) Schematic of high Double wishbone.

The curves that describe the variation of the characteristic angles, in the case of the high Double wishbone, tend to be more linear than the low Double wishbone. This is a positive feature because it guarantees progressivity to the dynamic performances of the vehicle. The limit of the high Double wishbone is the excessive size of the upright which increases the weight of the unsprung masses. For this reason, this scheme is mostly used in road cars.

The shock absorber was not shown in the previous images because, in the Double wishbone suspension, it has the only function of damper and spring support without affecting the kinematic mechanism. In the racing suspensions, in order to move the shock absorber spring assembly towards the center of the car and to make adjustments easier, *push rod* mechanisms are used. The following figure shows the visual comparison between a classic Double wishbone and a Double wishbone with *push rod*.

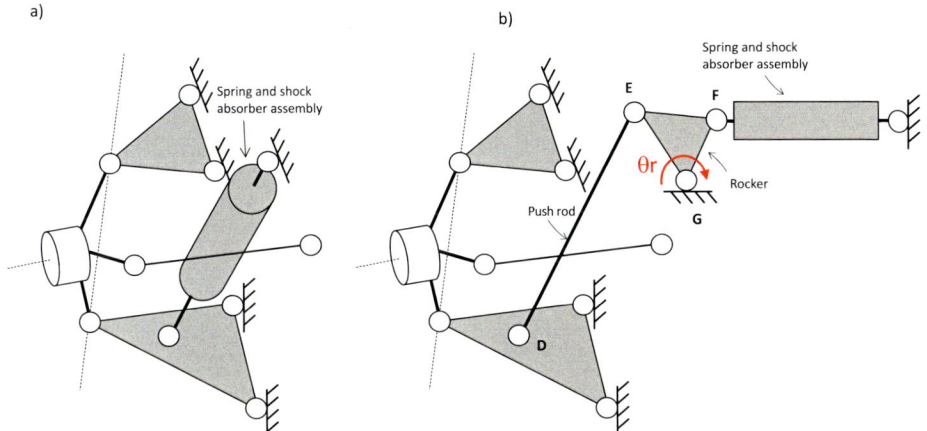

Figure 10 – a) Classic Double wishbone. b) Double wishbone with push rod.

In this mechanism, the rod *ED (push rod)*, pushed by the lower arm, moves the rocker which, in turn, pushes the shock absorber spring assembly into *F*. The ratio between the displacement of the point *F* and the displacement of the point *D* is the *Lambda* ratio and depends on the position of *D* and the measurements of the triangle EGF. The sizing of the rocker must be carried out in order to make the *Lambda* ratio constant during the excursion of the suspension. However, this goal is not easy to achieve. Any increase of this ratio can be exploited to produce a progressive rise in the slope of the suspension load-displacement curve. In this way, the load-displacement curve, like the one described at the end of chapter 3, could be obtained without the aid of the bump stop.

Moving the shock absorber spring assembly towards the center of the car lowers the yaw moment of inertia making the car more agile when changing direction. This is because the components approach the vehicle centerline and the sturdy and heavy structure, which must absorb the loads of the spring and shock absorber, is made in the center and not in the side area as in ordinary suspensions. Another scheme widely used in competitions is the *pull rod,* in which the rod does not work by compression but by traction. In this case, the sizing of the rod becomes easier to carry out because an element stressed by traction does not buckle, and therefore can have a smaller section, with a consequent gain in terms of weight. The following figure shows the schematization of the *pull rod* in which the rocker compresses the shock absorber spring assembly because it is pulled by the rod *DE (pull rod)* which, in turn, is constrained to the upright in *D*. It can be seen that this mechanism allows to lower the center of gravity of the suspension because, for how it is made, it provides for the arrangement of the shock absorber spring assembly in the lower part of the car.

Automotive suspension

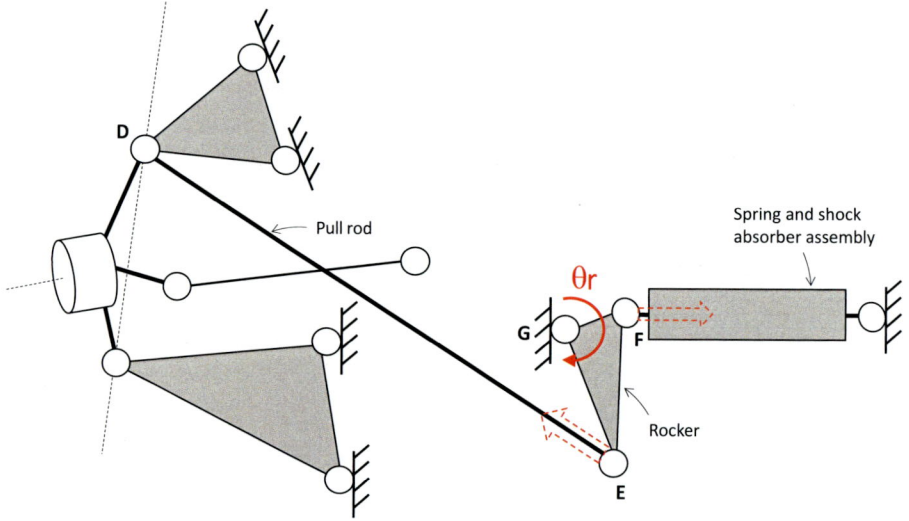

Figure 11 – Double wishbone with pull rod. In red the displacements of the points E and F of the rocker.

This last peculiarity, together with the possibility of making a lighter *rod* makes this mechanism advantageous from a performance point of view. However, it is disadvantageous, from a functional point of view, because it makes the adjustments related to the shock absorber spring assembly (preload and damping) difficult, since the latter is located in the lower part of the car in an area that is certainly not very accessible compared to the case of the *push rod*. Despite the importance of the reasons given, most of the time the choice between the two schemes depends above all on the aerodynamic constraints of the car which prefer one or the other solution based on the size of the rod and the shock absorber spring assembly towards the flow of air. For example, a *pull rod* scheme may not be suitable when the management of the air flow beneath the car is of fundamental importance for its performance.

5.2.2. McPherson

In the McPherson architecture the control of the angles of the wheel is guaranteed by control arm and telescopic shock absorber which, unlike to the Double wishbone, has the function of kinematic joint. The following figure shows, the schematization of the McPherson architecture with the shock absorber section (on the right) that participates in the kinematics as it imposes, on the two parts (the piston rod connected to the vehicle body by means of a ball joint and the strut housing fixed to the upright), the translation and rotation relative to the axis of the piston rod.

Suspension – Architectures and schematizations

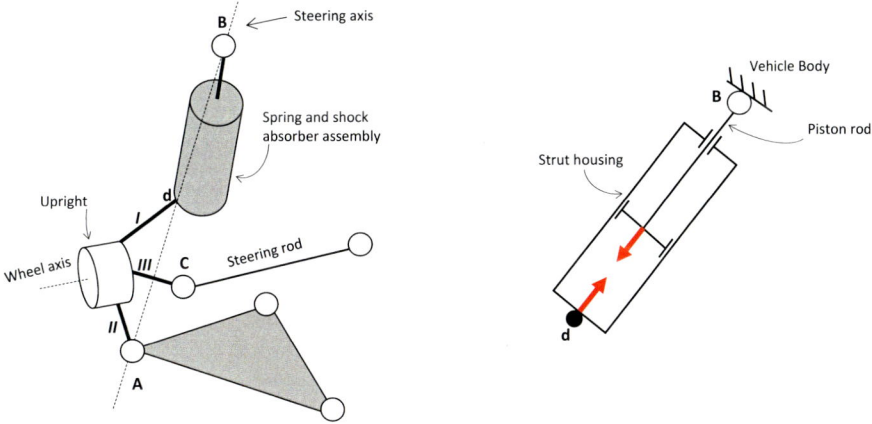

Figure 12 – Schematic of a McPherson with the section of the telescopic shock absorber.

The upright is a rigid body consisting of a cylindrical part on which the wheel bearing is keyed. It is rigidly connected to links *I*, *II* and *III* respectively connected to the shock absorber via two rigid fixings, to the arm and to the steering rod via two ball joints. The connection of the piston rod to the vehicle body can be made either with a ball joint, as happens for sports cars, or with a needle bearing or fifth wheel as happens for road cars. The last solution involves the interposition of an elastic block between the bearing (or fifth wheel) and the vehicle body, which has the task of filtering the road roughness and which gives to suspension greater *comfort* but less readiness.

The McPherson suspension fails to guarantee the same control of the characteristic angles (of the wheels) of the Double wishbone architecture. In particular:

- It has a limited *camber* gain both in bump travel (compression of the spring) and in steering and fails to guarantee the linearity of the variation of the characteristic angles of the wheel. For this reason, as we will see later, this architecture is suitable for cars with low *camber* gain on the rear axle in order to balance the two axles. The non-linearity of the variations in the characteristic angles gives the axle on which the suspension is installed a less progressive dynamic response than the Double wishbone.
- The lowering of the transverse inclination of the steer axis, a very sought after feature as will be seen later in the *KPI* study, is connected to the interference of the shock absorber toward the wheel because it is a consequence of the outward displacement of the point B.
- The absence of the upper arm, replaced by telescopic shock absorber, produces greater transverse flexibility which manifests itself with a greater *camber* loss under lateral load.

In conclusion, the McPherson is decidedly less performant than the Double wishbone but has more contained costs thanks to the lower number of components. For this reason, it is mostly used in small segments and in cars that do not have great dynamic performance.

By adding many complications to the classic scheme, it is possible to switch to the McPherson with revolving upright, which allows to lower the transverse inclination of steering axis and make the suspension more performant in steering. In the left part of the following figure the schematization of the assembly of this solution, while in the right part the exploded view are shown.

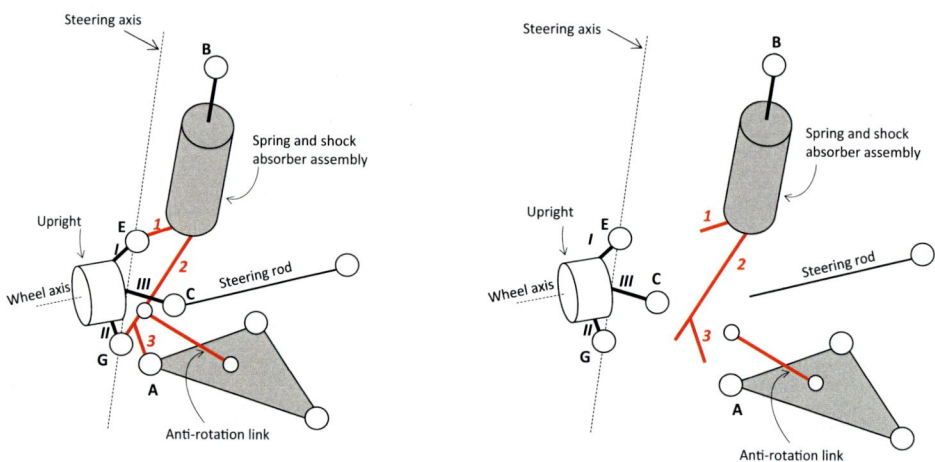

Figure 13 – McPherson with revolving upright.

The steering axis is obtained directly on the upright that rotates around the *EG* straight line thanks to the presence of the two-ball joints connecting the part consisting of links 1,2 and 3, rigidly connected to the shock absorber. In this solution, the presence of the anti-rotation link between the upright and the arm is essential to lock the steering of the upright around the straight line *AB*. In the following figure, the constructive difference between the new scheme and the classic one is shown. In the new scheme (left side figure) the upright is constrained to rotate around the two-ball joints thanks to which the new steering axis *b* is decidedly less inclined than the steering axis of the classic scheme *a* (figure right side) and closer to the center wheel. The solution described certainly brings performance improvements to the McPherson architecture but introduces construction complications and cost increases that are often not justifiable.

Suspension – Architectures and schematizations

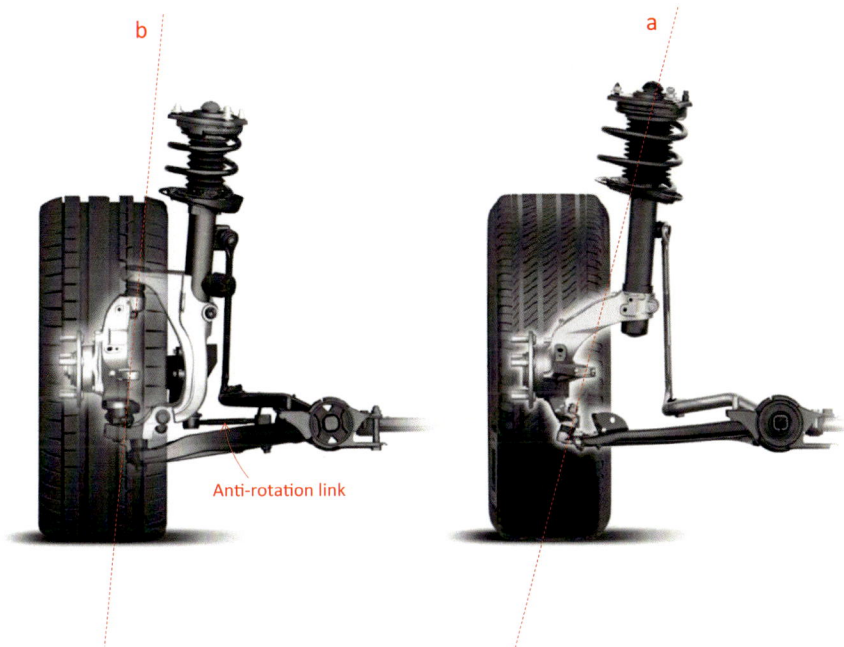

Figure 14 – Comparison between the positions of the steering axles of a classic McPherson (a) and of a McPherson with revolving upright (b) highlighting the necessary anti-rotation link between the arm and the body of the shock absorber.

5.3. Rear suspensions

The range of rear suspensions, unlike the front ones, is quite varied. Among the most used there are both independent wheel schemes and dependent wheel schemes. In this book we will describe only the most used architectures which include, for the dependent wheel architectures, the twist beam axle (or torsion beam axle), while for the independent wheel architectures the following schemes:

- Rear Double wishbone;
- Rear McPherson;
- Bilink;
- Multilink;
- Trailing arm;
- Semi-trailing arm.

The choice of the rear suspension architecture is always the result of a compromise between the car's performances needs, production costs and trunk volume. It is difficult to manage the compromise because it often happens that dynamically performing and inexpensive suspension architectures are very bulky towards the trunk.

Automotive suspension

5.3.1. Twist beam axle

The twist beam axle (or also twisting axle) is a type of rear suspension with dependent wheels mainly composed of a single axle and two anchoring bushings to the vehicle body. The central cross member consists of an open-section steel profile welded to the longitudinal arms on which the wheel hubs are screwed. The seats of the springs are obtained on shelves integral with the longitudinal arms and the shock absorbers are positioned externally to them to occupy as little space as possible the trunk. The elastokinematic behavior of this suspension depends on the torsional and flexural work of the central cross member and consequently on its shape. In particular, the position of the shear center of the central cross member section, at the vehicle centerline (see figure below), greatly influences the relationship between the *toe* variation and the *camber* variation in opposite wheel travel, as we will see later.

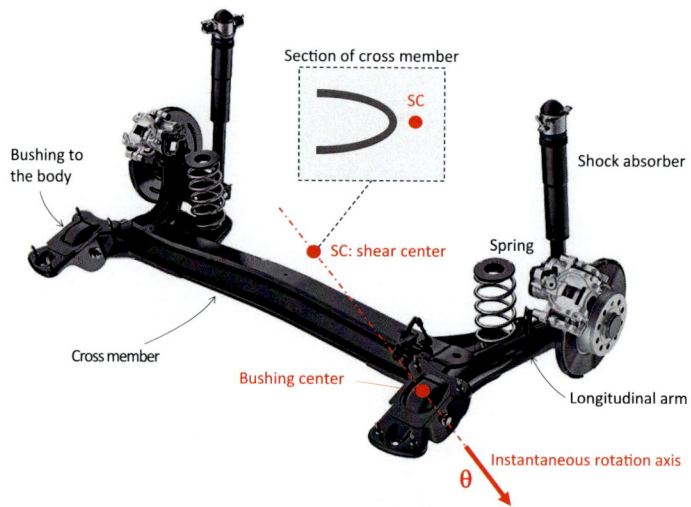

Figure 15 – Rear twist beam axle with the detail of the shear center in the section of central cross member at the center line. The figure also shows the construction of the instantaneous rotation axis.

The rear twist beam axle performances, whose comprehension will be clearer after reading chapter 6, is not the reason for choosing this architecture. The most obvious performance gap is the wheel opening under lateral load, which on a rear suspension causes the vehicle to oversteer and the consequent instability in cornering. This behavior is caused by the fact that the elastic bushing, which reacts to the lateral force (*Flat*), is advanced with respect to the latter and consequently the moment acting on the wheel tends to make it diverge, as can be seen from the plan view of the following figure. The only way to mitigate this defect is to induce the closure of the external wheel in opposite wheel travel so that balancing the undesired opening under load of the external wheel to the curve. To do this, it is necessary to position the bushing centers at a lower level than the shear center of the cross-member section relative to the vehicle centerline.

Suspension – Architectures and schematizations

This is because the instantaneous rotation axis (see chapter 6) of this suspension is, with good approximation[1], the junction between the bushing center and the shear center of the cross- member section, at the vehicle centerline (previous figure). In this configuration (shear center higher than bushing center), the decomposition along the x and z directions of the anticlockwise rotation θ around the instantaneous axis, induced by the suspension bump travel, produces wheel closure (the projection z is negative – downwards) and *camber* gain (the projection x is negative – towards the front axle), as shown in the lateral view of the following figure. This apparently complex concept will be clearer after understanding the functioning of the oblique arms and the axis of instant rotation, described in chapter 6. The *camber* gain, even if minimal, places the twist beam axle on a higher level, especially when it concerns high speeds, compared to those rear suspensions that have no *camber* gain.

Since the central cross member is always in an intermediate z between wheel center and bushing center, the optimal positioning of the shear center (*SC*) forces the wheel center (*WC*) to assume a greater height than the bushing center, as shown in the side view of the twist beam axle in the following figure.

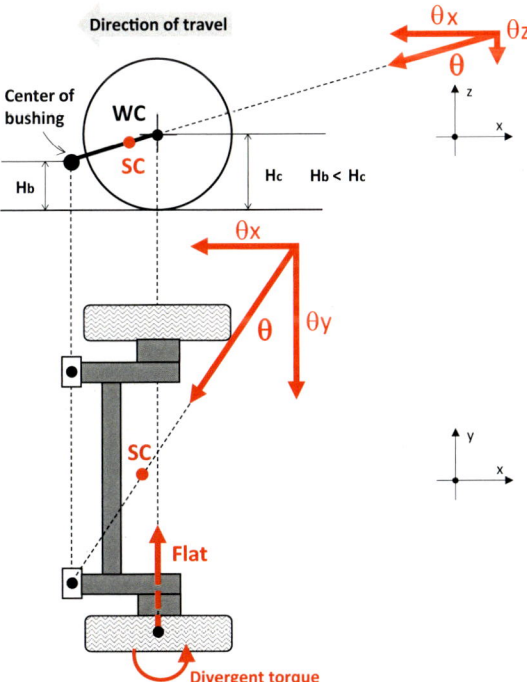

Figure 16 – Side view and plan view of the schematization of a twist beam axle. The projection of the rotation vector θ (translated in order not to complicate the drawing), on the x axis and on the z axis, respectively represent the *camber* gain and wheel closure. The divergent torque, due to the Flat, typical of this architecture, is highlighted in the plan view.

---

1. The approximation lies in the fact that this construction of the instantaneous rotation axis does not take into account the effect of the flexion of the arms and of the cross member which, until the FEM calculations are implemented, it is not possible to quantify.

It follows that the height from the ground of a car with a twist beam axle must never be increased by means of longer springs or spring seat spacers, as it would risk obtaining the wheel opening in opposite wheel travel, amplifying the undesired behavior under lateral load. On the contrary, by lowering the height of the car, a performance improvement of the suspension is obtained.

A second performance gap, due to the aforementioned positioning of the bushing center, is the advancement of the wheel center in bump travel, which significantly lowers the *comfort* of the car due to the wheel moving against the obstacle. The only way to contain this unpleasant effect is to lower the longitudinal stiffness of the elastic bushing, taking into account the consequences on lateral dynamics.

In conclusion, we can say that, if well designed, the twist beam axle in addition to having an excellent ratio between lateral dynamic performance and construction costs, allows to increase the volume of the trunk because it has a very small size. For this reason, many car manufacturers adopt this solution not only for small cars but also for higher segments of the market. However, due to the transverse dimensions, the classic twist beam axle does not adapt to rear-wheel drive and the presence of additional components, as in CNG cars. In the presence of rear-wheel drive, arched twist beam axle schemes can be adopted which have a shaped central cross member to allow the rear differential to be housed.

Despite the simplicity of construction, the design of the twist beam axle is very complex because to correctly size the shape of the central cross member and the arms it is necessary to use FEM (finite element method) calculation algorithms, the use of which requires a significant professional experience.

5.3.2. Rear Double wishbone and rear McPherson

Double wishbone and McPherson installed on the rear axle do not have the overexposed limits to achieving the desired elastokinematic performances. The designer, by arranging the arms and the steering rod in the suitable way, almost always obtains the target curves. Generally, in the rear Double wishbone, to obtain wheel closure under lateral load (desired performance on the rear, as will be clarified in chapters 6 and 8), point *A* (following figure) is set back with respect to point *B* so that the moment of the force at ground makes the wheel converge towards the center line of the vehicle. In the following figure, the side view of a rear Double wishbone with the *Flat* lateral force (towards the inside of the vehicle) which, due to the way the steering axis is positioned, produces the moment *M* which tends to close the wheel (*toe-in*).

As with the front axle, the Mcpherson has greater limitations relative to the *camber* gain during the vertical motion of the wheel. For this reason, since this limit induces the vehicle to undesirable behaviors (major tendency to oversteer and lateral response delays), the more performant cars adopt Double wishbone or Multilink architectures. In some situations, since the elastokinematic characteristics pursued depend on those present on the other axle, if on the front axle there is a not performant suspension, the use of a rear Double wishbone becomes useless if not deleterious.

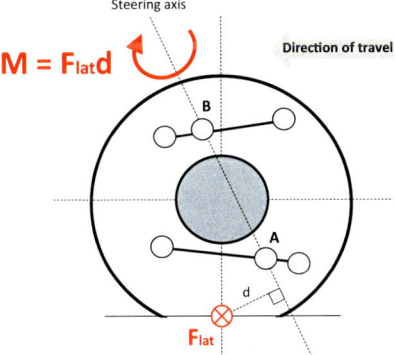

Figure 17 – Side view of the kinematic scheme of a rear left Double wishbone in which point A is set back with respect to point B and the moment of the lateral force, which penetrates the figure, tends to close the wheel.

With regard to the trunk, the McPherson suspension is much bulkier than the Double wishbone one due to the presence of the spring-shock absorber group which, having an important size, requires a certain transverse inclination in order not to interfere with the wheel.

5.3.3. Bilink

The bilink independent wheel suspension is an excellent compromise between costs, performance and trunk volume. In the figure an example:

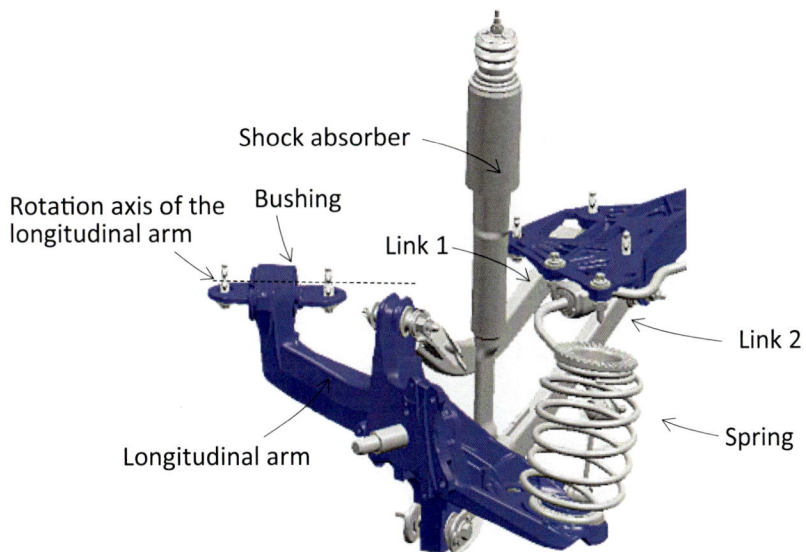

Figure 18 – Bilink rear suspension.

The performances are guaranteed by the fact that the variation of the position of the three links produces large margins of variability of the kinematic characteristics. The costs are guaranteed by the limited number of components and the reduced sizes are guaranteed by the positioning of the spring and shock absorber similar to what happens on the twist beam axle. It is an excellent solution for discrete performances (not at the level of Double wishbone and Multilink), a spacious trunk and not excessive costs. For this reason, it is the rear suspension adopted by most of the sporting compact C-segment because it allows good performance without sacrificing too much space that is scarce in these cars. Similarly, to the twist beam axle, in order to have the desired wheel closure in bump travel of the wheel (compression of spring), it is necessary that the intersection between the straight lines (joining the joint centers) of links 1 and 2 has a greater height than the bushing center of the longitudinal arm. The more this height increases, the more the ratio between wheel closing and *camber* gain in bump travel increases. This is because the axis of instantaneous rotation of the wheel (chapter 6) is approximately[2] the junction between the bushing center and the intersection between the lines of the two links. If the lines of the two links are skewed, the midpoint of the segment of minimum distance shall be considered.

5.3.4. Multilink

The rear Multilink is an architecture consisting of a certain number of links and other components which, appropriately arranged, offer excellent margins for handling the wheel. Bilink also belongs to this family, but it was treated separately, due to the low number of components. The increase in the number of links, necessary to improve the performance level of the suspension, increases the design and production costs. For these reasons, this architecture is used on very expensive and performing cars. This paragraph will only deal with the conceptual aspects of this type of suspension without going into the details of the various applications.

In the following figure an example of Multilink obtained by replacing two links to each arm of a Double wishbone is shown. The upright consists of the cylindrical part on which links I, II, III and IV are rigidly connected. In the case of a front Multilink, the steering axis can be identified by joining the intersection points of the extensions of the two upper links (link1 and link2) and the two lower links (link3 and link4). If the straight lines are skewed, the midpoint of the minimum distance segment is considered. In this type of suspension, the steering axis can assume any position because it is not linked to joints *A* and *B*, whose displacement margins depend on interference with the wheel. This, as will be seen in the next chapters, represents a big advantage because it allows you better manage the steering geometry.

The performance increase that the rear Multilink guarantees mainly concerns the vari-

---

2. The approximation lies in the fact that, generally, the axis of the longitudinal arm bushing does not meet the junction of the two transverse links, but its inclination is determined in the elastokinematic calculations as a function of the wheel closure objective under lateral load.

ations under load of the characteristic angles. This is because, by moving the links so as not to vary the position of the axis of instant rotation of the suspension (see chapter 6), the elastic displacement relative to some bushing can be enhanced or inhibited. For example, if link3 is moved forward on the left rear suspension in the following figure, along the plane containing link3 and link4, the closure (*toe-in*) under lateral load decreases without modifying the slope of the *camber* and *toe* trends in bump travel. This concept will certainly become clearer after reading chapter 6.

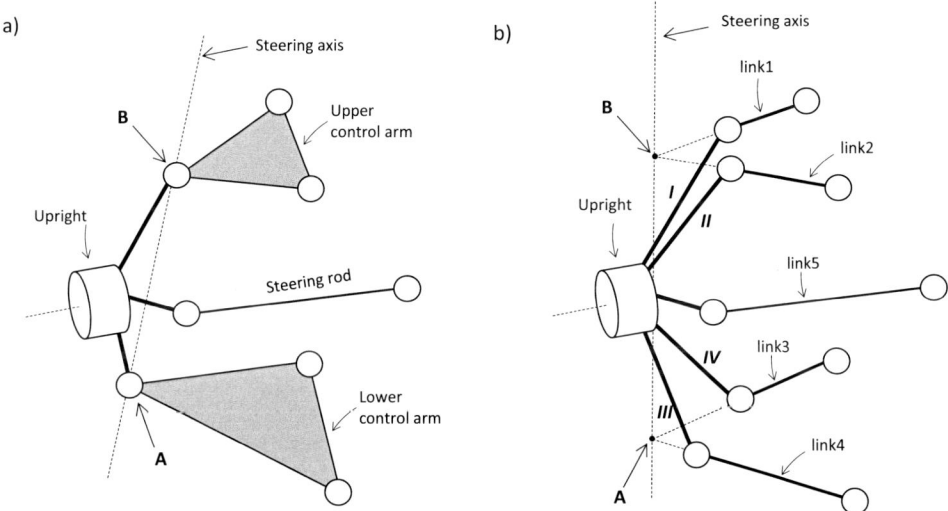

Figure 19 – a) Rear left Double wishbone b) Rear left Multilink.

Using this technique in positioning the arms (keeping the position of the instantaneous rotation axis constant), which will be clearer after reading chapter 6, it is possible to further improve the performance under load by inserting other links that inhibit unwanted elastic deformation. Obviously, it must be taken into account that these actions increase both the costs related to the product (materials, design, etc.) and those related to the industrial production, given the greater number of components to be managed in the assembly line.

5.3.5. Trailing arm and semi-trailing arm

In the longitudinal arm scheme, the upright is integral with an arm that has the rotation axis parallel to y, as can be seen in the following figure which shows the side and plan view of the architecture. From the figure it is clear that the position of the rotation axis affects the displacement of the point on the ground. The figure also shows the rotation θ around the rotation axis both as an angle and as a vector.

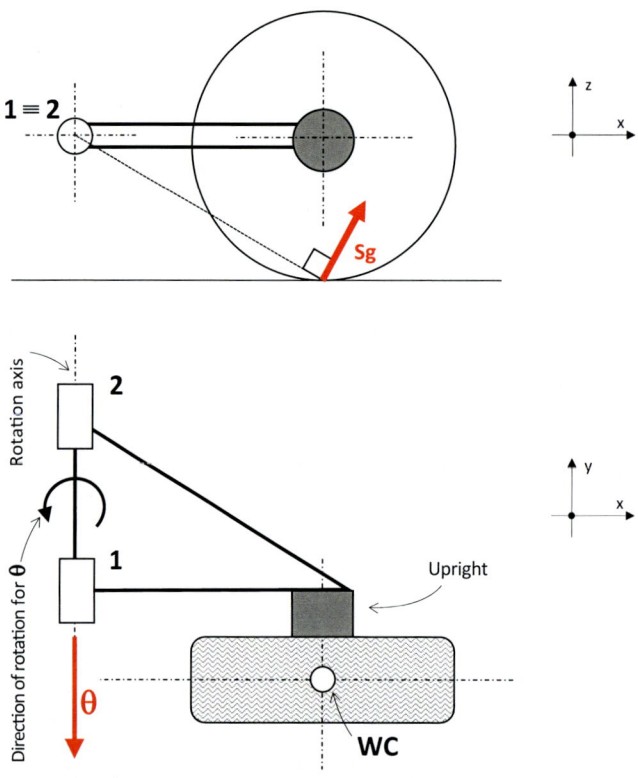

Figure 20 – Trailing arm. Side view and plan view.

In this type of suspension, which boasts a great constructive ease, the *camber* gain and the *toe* variation in bump travel (compression of the spring) are absent because the rotation vector has no projection on the x axis and on the z axis, which respectively represent the variation of *camber* and convergence. For the aforementioned reasons, this architecture is used on low-cost vehicles, without great performance ambitions and with a large trunk. When the rotation axis is not parallel to the transverse axis, the scheme becomes oblique arms and is called semi-trailing arm. In particular, if the rotation axis is tilted by ΔX between the bushing centers 1 and 2, as shown in the following figure, we see that the θ rotation vector also acquires the component θx. This, being the rotation of the wheel around the longitudinal axis, represents the desired *camber* gain.

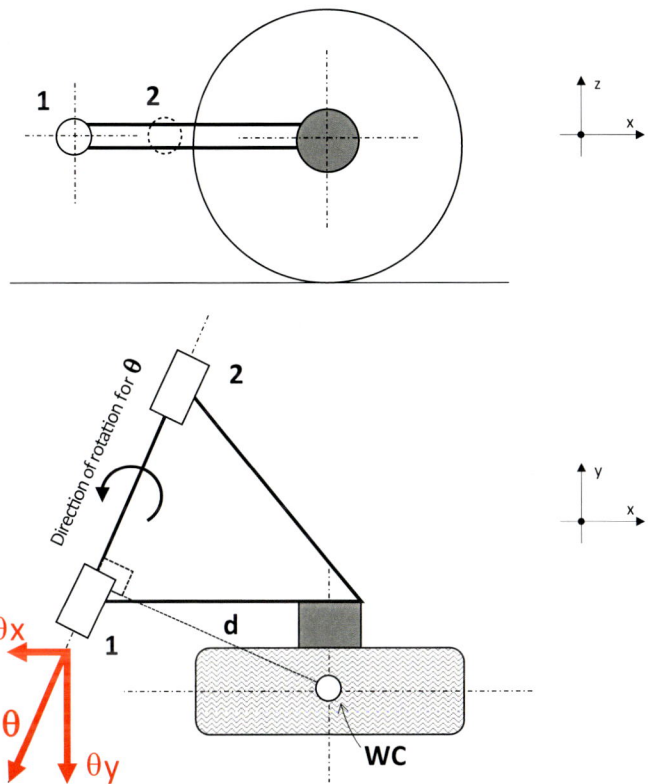

Figure 21 – Semi-trailing arm with only *camber* gain. Side view and plan view.

If the sizes of the arm installed are known, it is possible to calculate the slope of the *camber* variation as a function of the bump travel (dz) by using the following relationship:

$$\frac{Delta\ Camber}{dz} = -\frac{(X2 - X1)}{D(12)}\frac{1}{d}$$

where $D(12)$ is the distance between the center of the bushing 1 and the center of the bushing 2 and d is the distance between the center of the wheel (always located in the center of the tire) and the straight line 12. Since these last two quantities are always positive, in order to have *camber* gain, it is necessary that $X2 > X1$ and thus that point 2 is further back than point 1.

Automotive suspension

Similarly, to obtain a *toe* variation in parallel travel of the wheel, it is necessary to introduce a ΔZ between the bushing centers 1 and 2, as shown in the following figure:

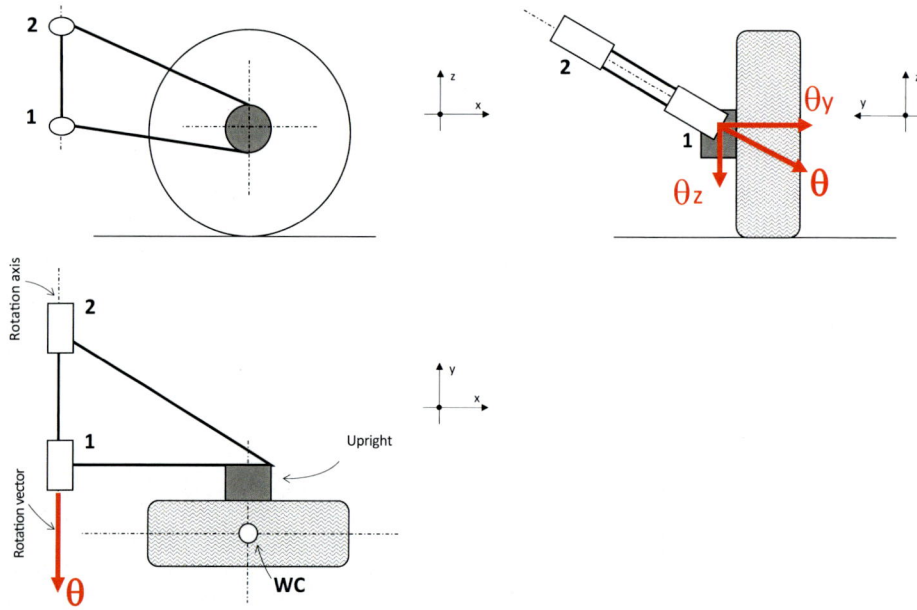

Figure 22 – Orthogonal projections of a Semi–trailing arm suspension with the only closing wheel in bump travel.

It is clear that the inclination assumed by the rotation axis brings out the θz component which represents the rotation of the wheel around the vertical axis and therefore the *toe* variation. In particular, if point 2 is higher than point 1, the wheel closes in bump travel (compression of the spring). The slope of *toe* variation is as follows:

$$\frac{Delta\ Conv}{dz} = \frac{(Z2 - Z1)}{|Y2 - Y1|} \cdot \frac{1}{(Xcr - X1)}$$

where the difference $Y2-Y1$ is expressed as an absolute value to simplify the application of the formula. Since Xcr (x wheel center) is always greater than $X1$, the wheel closes bump travel when $Z2 > Z1$.

Condensing the two trends, it is possible to create an oblique arm (semi-trailing arm) with variation of *camber* and *toe* in bump travel by applying both a ΔX and a ΔZ between points 1 and 2, as shown in the following figure.

Suspension – Architectures and schematizations

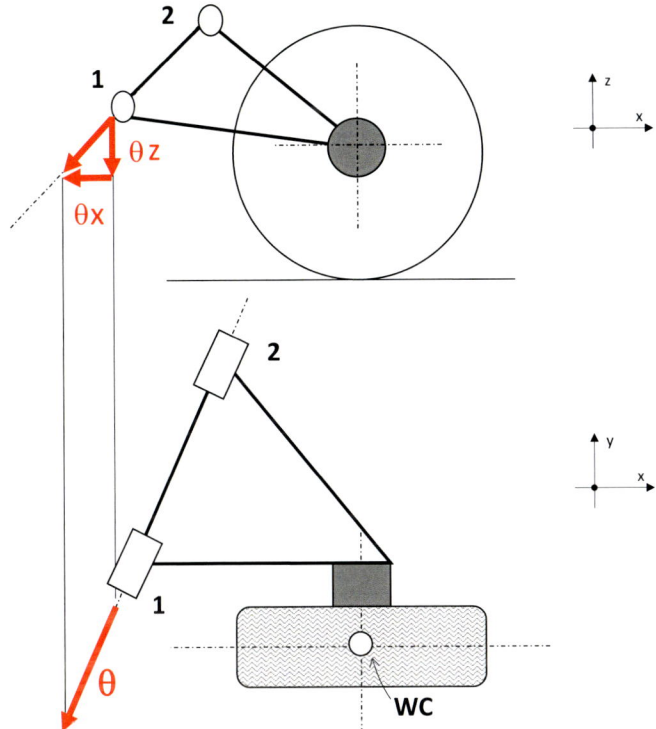

Figure 23 – Semi-trailing arm with *camber* gain and positive *toe* variation in bump travel. Side view and plan view.

With this arrangement of the points, the ratio between the *toe* variation and the *camber* variation in bump travel is:

$$\frac{Delta\ Conv}{Delta\ Camber} = -\frac{(Z2-Z1)}{(X2-X1)}$$

Considering the wheel closure positive and *camber* gain negative, the ratio will be negative when the wheel closes and gains *camber*.

The important *camber* gain that these suspensions are able to guarantee is not always a positive aspect, especially at high speed in extreme conditions. In fact, when the tire is around its peak of maximum lateral force, small vertical displacements of the wheels induced by any external perturbations correspond to large variations in *camber* and lateral sliding which can cause sudden loss of adherence and consequent oversteer. This is because the increase in *camber* always corresponds to a decrease in the *sliding limit*. For high speeds, a more progressive suspension is recommended, which produces less variability in the maximum lateral forces, and makes the vehicle more stable and controllable in extreme conditions.

6. SUSPENSION
Set-up and operation

6.1. General description

The suspensions are created to isolate the body of the vehicle from the stresses deriving from the roughness of the road surface, in order to ensure the occupants a certain level of *comfort*. Isolating means appropriately lowering the values of the peaks of the forces that are transmitted to the vehicle body and dissipating the kinetic energy acquired by the masses, as quickly as possible, to avoid the triggering of annoying vibratory phenomena. In addition to isolation, the suspensions also perform the task of leading the relative motion of the wheels with respect to the vehicle body to ensure good *handling* for the vehicle. *Handling* represents the ability of the vehicle to deal with the trajectories that are imposed on it and the ease with which the user can do it. *Handling* depends mainly on the lateral dynamic behavior of the vehicle, that is, on the trend of transverse forces and slip angles that arise from a change in trajectory. The main difficulty in designing a suspension lies in the fact that the pro-comfort direction and the pro-handling direction are generally conflicting. In fact, a vehicle that is fast and ready to change direction produces, with the same steering wheel angle imposed, very abrupt, annoying and therefore not very comfortable lateral acceleration. Similarly, a vehicle with very soft suspension, that filters the road surface well, will be slower in changes of direction because, at the set steering wheel angle, the greater roll angle (due to the flexibility of the suspension) delays the lateral response.

The suspensions determine the relative positions of the wheels with respect to the vehicle body. The static position is that assumed by the wheel when the car is not moving. On the other hand, the dynamic positions are those assumed by the wheel when the car is moving, when the suspension carries out the motions permitted by its architecture, which can be assimilated mainly to:

1. Bump-Rebound motion: It consists of the vertical displacement, in positive z (bump) and in negative z (rebound), of the wheel center while keeping the vehicle body locked.
2. Steering: It consists of the rotation of the wheels around the respective steering axes.

The dynamic positions also include those assumed by the wheel under the various forces deriving from the operation of the car, which depend on the flexibility of the elastic bushings that connect the suspension components and the vehicle body.

6.2. The characteristic angles of the wheels and the geometric quantities of the suspension

The *characteristic angles* are the inclinations that define the orientation of the wheel with respect to the vehicle body, both in static and dynamic conditions. The *geometric quantities* are measures that depend on the geometry of the suspensions and which, together with the *characteristic angles*, vary according to the forces and displacements due to the operating conditions. It is called elastokinematics of suspensions, the correlation between the variations of the aforementioned quantities and the causes that trigger them. The study of elastokinematics of the suspensions deals with monitoring the trend of the *characteristic angles* and the *geometric quantities*, during some standardized maneuvers that simulate the operation of the suspensions under different operating conditions. The maneuvers are as follows:

1. Parallel wheel travel: the vehicle body is locked and the wheels move in phase in the vertical direction. This maneuver is used to simulate the changes in vehicle height due to different load conditions and to calculate the bump stiffness of the axle.
2. Opposite wheel travel: the vehicle body is locked and the wheels move according to the vertical direction in phase opposition. This is used to evaluate the trend of the *characteristic angles* and the *quantities* in the curve regardless of the lateral loads and to calculate the roll stiffness of the axle.
3. Steering: the wheels steer around the corresponding steering axes, controlled by the steering mechanism. This maneuver serves both to monitor the trend of the *characteristic angles* and *geometric quantities* and to evaluate the relative steer angles of the wheels to identify the *Ackermann* percentage.
4. Lateral load at ground: application of a lateral force in *y* which simulates the reaction to the centrifugal force acting on the wheel outside the curve and which, due to the compliance of the suspension attachments on the vehicle body, induces the wheel to move according to the six degrees of freedom. This maneuver is used to evaluate the trend of the *characteristic angles* and of the *geometric quantities* in the curve regardless of the opposite wheel travel.
5. Longitudinal load on the ground: application of a longitudinal force at ground in *x* directed towards the rear of the car that simulates braking on each of the wheels.
6. Longitudinal load at wheel center: application of a longitudinal force at wheel center in *x* directed forward to simulate the drive torque and directed towards the rear to simulate the passage of the obstacle.

Suspension – Set-up and operation

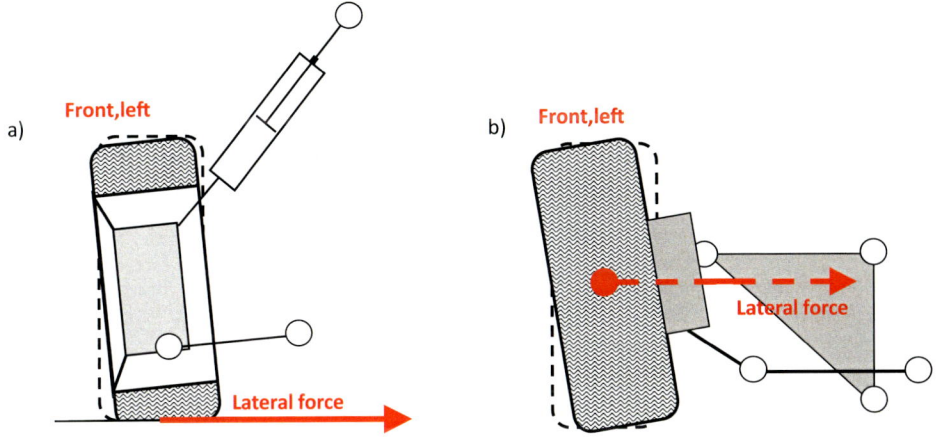

Figure 1 – View from the rear of the car (a) and plan view (b) of the left front wheel loaded with a lateral force towards the inside of the vehicle. In view (a) you can see the loss of *camber* and in view (b) you can see the *toe-out* of the wheel.

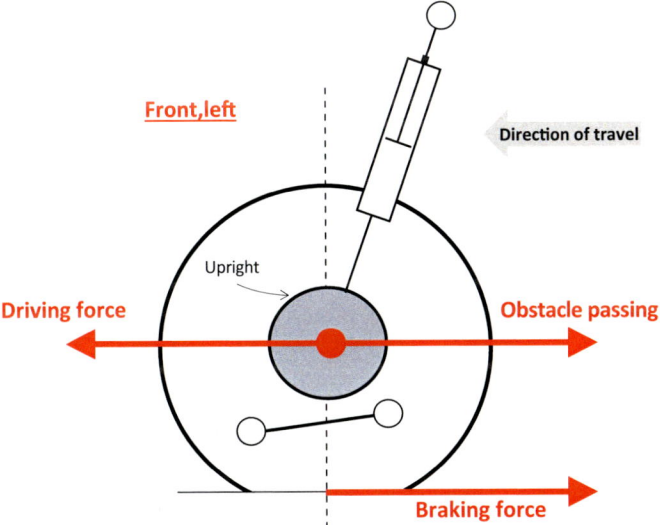

Figure 2 – Side view of the left front suspension with the longitudinal forces of braking, acceleration (in the case of drive shaft) and obstacle passing.

The characteristic angles monitored are:
- · Camber;
- · Toe.

The geometric quantities monitored are:
- · Caster;
- · King Pin Inclination;
- · Scrub radius;
- · Caster trail;
- · KPIoffset;

- *Steering axis offset (SAO);*
- *Wheel track at ground;*
- *Roll center;*
- *Anti dive/lift;*
- *Distance between the front wheel center and a point on the vehicle body;*
- *Distance between the rear wheel center and a point on the vehicle body.*

The first six *geometric quantities* concern the steering system and will be described in the next chapter. The last two quantities, summed to the radii under load, define the height of the car from the ground.

The aforesaid quantities, if measured with the car stationary, represent the static characteristic angles and the static geometric quantities. To define them, we must establish a vehicle load condition because they depend on how the suspension is compressed due to the weight on the car. The load condition that is generally considered is that obtained with the car in running order (with all auxiliary liquids), with a full tank of fuel and without any occupant. For sports cars it is advisable to consider the previous condition plus the weight of the driver.

In the following paragraphs the *characteristic angles* and the *geometric quantities* will be described with reference to both the static condition and the dynamic condition.

6.3. The *camber*

The *camber* angle is the angle that the longitudinal plane of the tire forms with the vertical. It is negative when the upper part of the tire is closer to the centerline of the vehicle than the lower part.

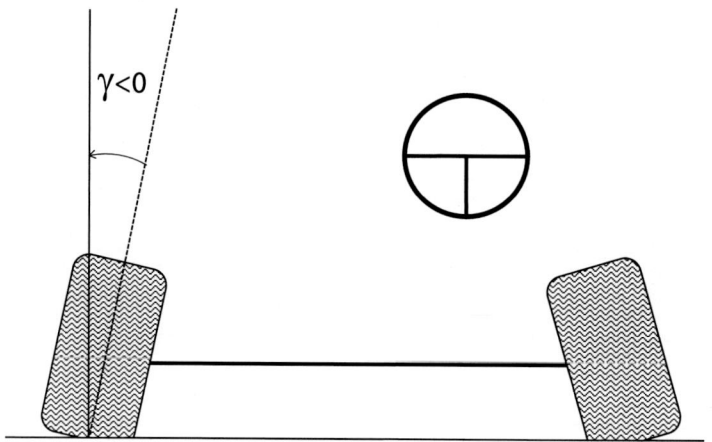

Figure 3 – Front view of the front axle. *Camber* angle <0.

We will only consider negative static *camber* because, except in very few cases that we will not deal with, there are no cars that adopt positive values. For convenience, when comparing two *camber* angles, we will assume that the greater is the one with the highest absolute value. Negative values are preferred because, as seen in chapter 2, with the same transverse force, they lower the slip angle of the axle. Consequently, a more "cambered" axle undergoes an increase in slip stiffness. It is thus clear that, by distributing the *camber* angles between the axles, the slope of the understeer curve can be increased or decreased, since *Kus* (see chapter_4) is linked to the difference between the front slip stiffness and the rear slip stiffness. Increasing the *camber* only on the rear axle increases the slope of the understeer curve and the vehicle becomes " less direct ". On the other hand, increasing the *camber* on the front decreases the slope of the understeer curve and the vehicle becomes " more direct ". Therefore, by adjusting the difference between front and rear *camber* a car can become more or less direct. The following figure shows the trend of the slope of the understeer curve as the front *camber* increases. The figure focuses only on the slope, however, the front *camber* also influences, not significantly, the extension of the linear section of the understeer curve, as it helps to shift the *sliding limit* of the axle characteristic. This will be more clear in chapter 8.

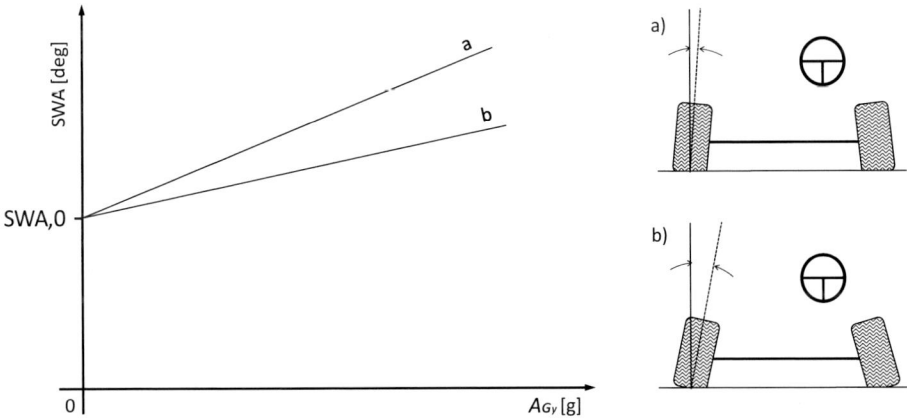

Figure 4 –Example of variation of the understeer curve as the front *camber* varies. Curve b, more direct, is the one corresponding to the axle with more *camber*.

Generally, in road cars the static *camber* values on the front wheels range from -0.5 to -1.0 degrees while on the rear wheels from -0.5 to 1.5 degrees. The minimum values correspond to small cars, while the maximum values correspond to the sportiest cars. Road cars adopt greater static *camber* angles on the rear axle to increase the slope of the understeer curve and make the cars understeer more. In racing cars, the values not only increase but overturn, meaning that the front axles have *camber* angles that can exceed -3 degrees and are greater than the angles on the rear axles. The reason for this choice lies in the fact that a greater static *camber* on the front axle compared to the rear makes the car more direct, more precise and faster when changing direction. These characteristics, fundamental in racing cars, are not required in road cars because an inexperienced user would not be able to manage a car that is very direct and very quick in changing directions.

The value of the *camber* angle, due to the movement of the suspension, undergoes variations led by the kinematics and the stiffness of the connections between the suspension and the vehicle body. In particular, the roll of the car induces the axles to opposite wheel travel and the consequent variation of the characteristic angles. Taking as reference only the external wheels, which always have a greater weight for cornering dynamics as they absorb a higher percentage of lateral force, it is very important that the increase in the front *camber* is well balanced with the increase in the rear *camber*. The balance between these two defines the stabilization of the vehicle after the transient cornering as described below.

In general, during a curve maneuver, due to the load transfers and to the variation of the characteristic angles due to the roll, it is necessary to correct the steering wheel angle set initially. The steering wheel angle necessary to face the transient and the stabilized section of a curve starts from a value ϕ and is increased by a $\Delta\phi$ (figure below) during the interval in which the outer wheels bump due to the roll and changing their orientation and the vertical load. The ideal condition is that $\Delta\phi$ is concordant with the initial ϕ so that the steering angle is increased.

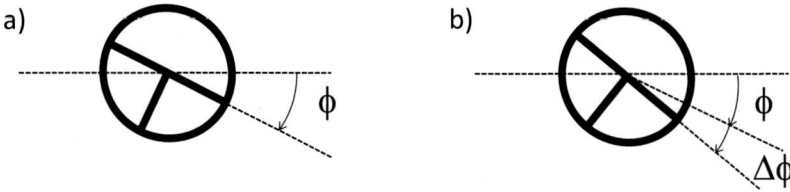

Figure 5 – a) Steering wheel angle in corner entry, b) Steering wheel angle in the stabilized section.

The sign and the value of the $\Delta\phi$ depend on the variation of the characteristic front and rear angles which in turn depend on the forces deriving from the curve. We shall now develop the dependence of $\Delta\phi$ on the *camber* by referring to the following paragraphs the dependence on the other angles.

Regarding the variation of *camber* we can say that the $\Delta\phi$ is concordant with ϕ if the rear gain (in bump travel) exceeds the front one. In this way the slip stiffness of the rear axle exceeds that of the front axle and consequently increases the difference $(\alpha_f - \alpha_r)$ (chapter 4) which makes it necessary to increase the steering angle to stay on the trajectory. If, on the other hand, the front *camber* gain exceeded the rear one, the car would become more direct during the evolution of the curve and consequently (considering only the influence of the *camber* and not of the other angles) there would be a negative $\Delta\phi$, which would induce the user to counter-steer. The latter behavior is undesirable because it is annoying, unstable and difficult to manage. Below is an example of front *camber* and rear *camber* trend for a competition vehicle of the 2000cc prototype class.

Suspension – Set-up and operation

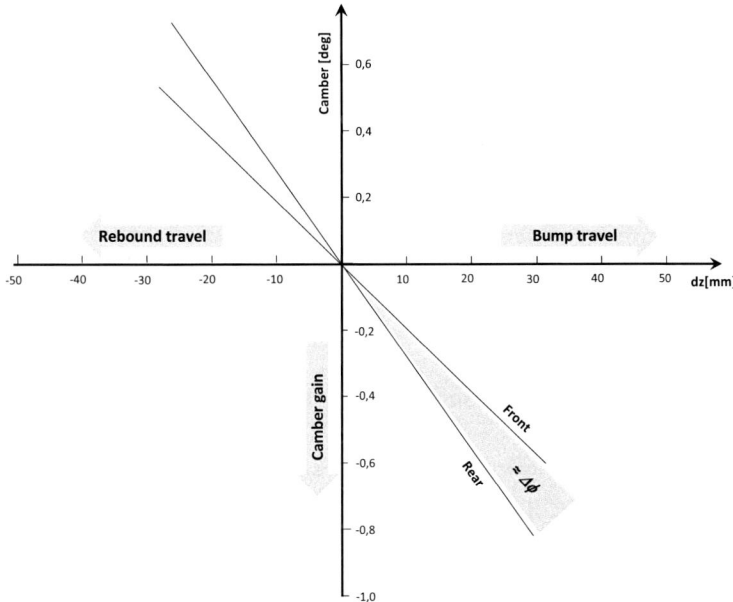

Figure 6 – Variation of front and rear *camber* in parallel wheel travel with evidence of the difference between rear and front, which is proportional to $\Delta\phi$.

With the same bump travel (compressed suspension) of the outer wheels, the rear *camber* gain must be greater than the front gain. The stabilized $\Delta\phi$ also depends on the variation of *camber* as a function of the lateral load on the ground resulting from the lateral acceleration of the vehicle. In particular, for the $\Delta\phi$ to increase, the loss of front *camber* must be greater than the loss of rear *camber*. Below is the desired trend of *camber* variations under lateral load.

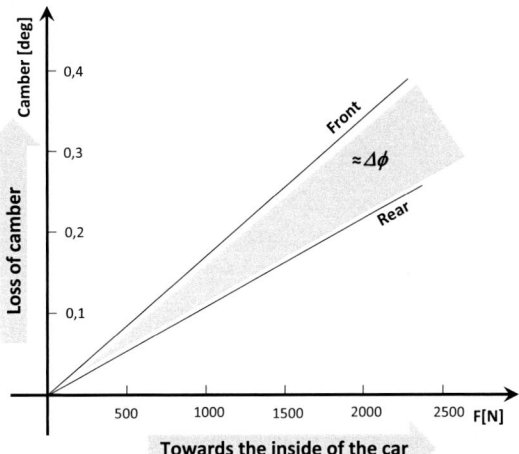

Figure 7 – Variation of front and rear *camber* under lateral load with evidence of the difference between rear and front affecting the $\Delta\phi$.

Automotive suspension

In racing cars the goal is to have the least loss of *camber* under lateral load because the correction must be as small as possible, as correcting means losing time. The loss of *camber* depends on the elastic deformations of the vehicle body and of the elastic bushings. They are necessary elements to give greater *comfort* to a suspension because they filter the transmission of forces to the vehicle body. For this reason, they are always present in road cars but, since they produce response delays, they are absent in racing cars where very rigid ball joints are used.

In conclusion, the differences between the curves of the two previous figures are proportional to the $\Delta\phi$. The designer must appropriately size these differences, in order to obtain the desired $\Delta\phi$.

The correct sizing of the balance between the *camber* angles of the axles has a very important weight on the dynamic performance of the vehicle. The trend is to adopt more generous *camber* angles to improve vehicle response. The limits of this trend are as follows:
- Tire wear;
- Resistance to advancement;
- Decrease in the adherence capital under braking and acceleration;
- Uniformity of the tread temperature;
- Decrease of the *sliding limit*.

In chapter 2 a more detailed description.

6.4. The *Toe*

The *toe* angle or convergence angle is the angle that the trace – on the horizontal plane – of the longitudinal plane of the tire forms with the longitudinal plane of the vehicle. It is generally measured in degrees but sometimes also in millimeters and is positive when the wheel tends to converge towards the direction of travel of the vehicle. The measurement in millimeters is the difference between the distances d2 and d1 between the extremes of the two rims, evaluated at the same height of the wheel center, as illustrated in the right part of the following figure. This measure defines the *total toe* and it is widely used to make adjustments where suitable equipment is not available, such as on the track.

The asymmetry of the front static *toe* angles is the main cause of the misalignment of the steering wheel. When the front *toe* $\varepsilon_{f,l}$ and $\varepsilon_{f,r}$ are different, for the vehicle to proceed in a straight direction, it is necessary to rotate the steering wheel angle by an angle δ_s. The following formulation is very effective for calculating the angle δ_s in degrees, starting from the measurements of the front *toe* in primes.

$$\delta_s = \frac{(\varepsilon_{f,l} - \varepsilon_{f,r})}{2} \frac{\tau}{60}$$

Remembering that τ is the steering ratio, that is the ratio between the steering wheel angle and the average of the wheel angles, considering the clockwise steering wheel angles as negative. From the formula, approximated because it does take into account the rear *toe* for the following reasons, it can be seen that, in a car with a steering ratio equal to 15, a difference of only 8 primes between the two wheels (left wheel opened by 4 primes and right wheel closed by the same amount) produces a misalignment of one degree. In the left side of the figure 9 an example is shown in which the left convergence angle is different from 0 while the right is equal to 0. The steering wheel angle that allows the car to travel straight is the one that makes the *toe* of the two wheels symmetrical (right side of the figure 9).

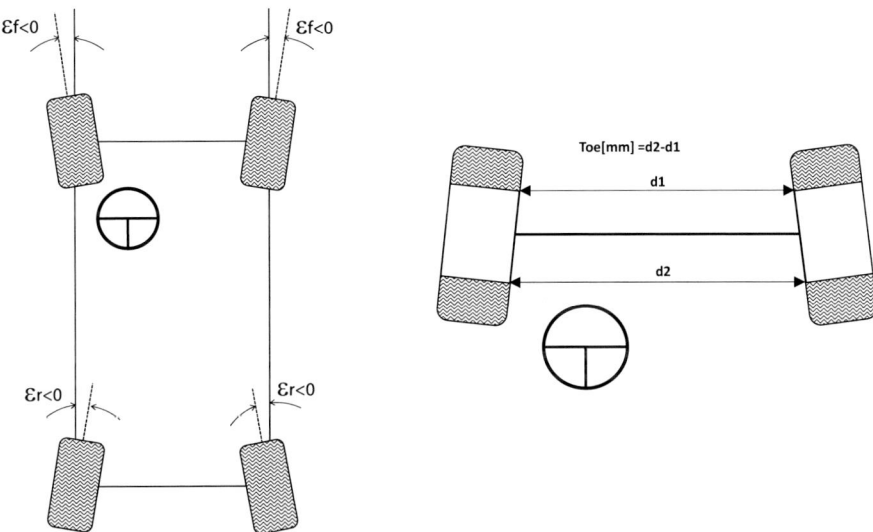

Figure 8 – *Toe* angle expressed in degrees and in mm. Plan view.

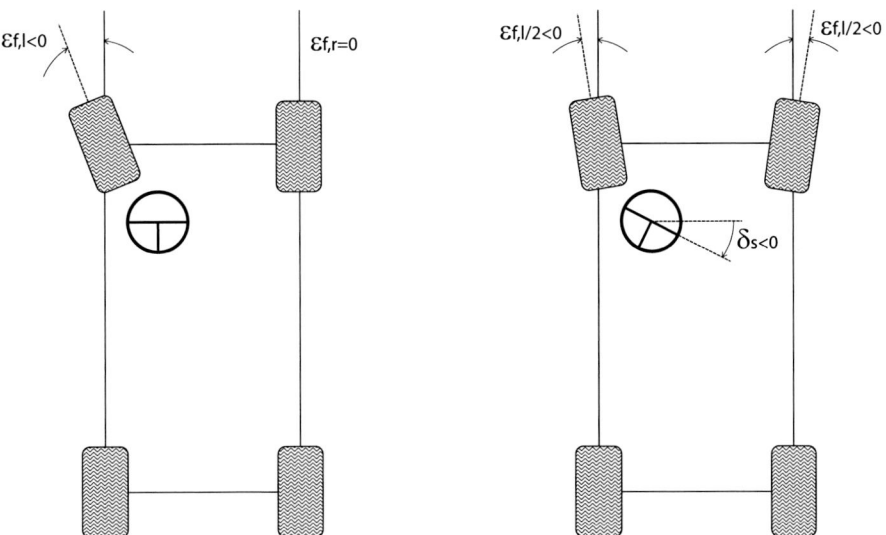

Figure 9 –The steering wheel angle that allows straight travel must make the front *toe* symmetrical.

The asymmetry evaluated on the rear axle produces what is called the *thrust axis*, represented in the following figure. With the same asymmetry, the rear axle causes less steering wheel misalignment with respect to the front axle. This is because the resistance force to advance (from 100 to 400N approximately on 1000kg depending on the type of tire and the test speed), in the presence of a caster angle other than zero (see next chapter), tends to realign the wheels, thanks to the presence of the elasticity of the bushings. This realigning effect does not occur in those rear suspensions which, to favor wheel closure under lateral load (see next paragraph), have a negative *caster*. For these reasons, in the presence of elastic bushings and positive *caster*, the small asymmetries on the rear can be neglected in the previous formula. This does not happen on the front axle which has labile wheels when steering, and all the stresses deriving from the asymmetry are absorbed by the free rotation of the pinion, which is transferred to the steering wheel without any impediment.

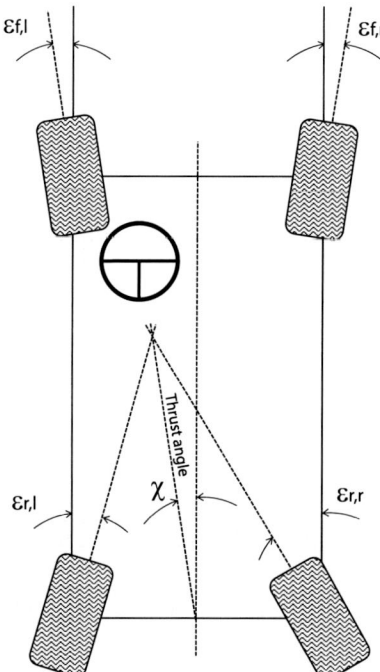

Figure 10 – Thrust angle construction.

To motivate the static *toe* set-up it is necessary to make some assessments on the effect on the vehicle dynamics. We can say that the static *toe-out* on the front axle increases the slope of the understeer curve (chapter 4) because, in stabilized condition and with the same trajectory, a greater steering angle is required the more the outer wheel is open. On the other hand, the front *toe* has no influence on the sideslip curve because the latter depends only on the *toe* of the rear axle. In the following figure the summary of what has been said is presented.

Suspension – Set-up and operation

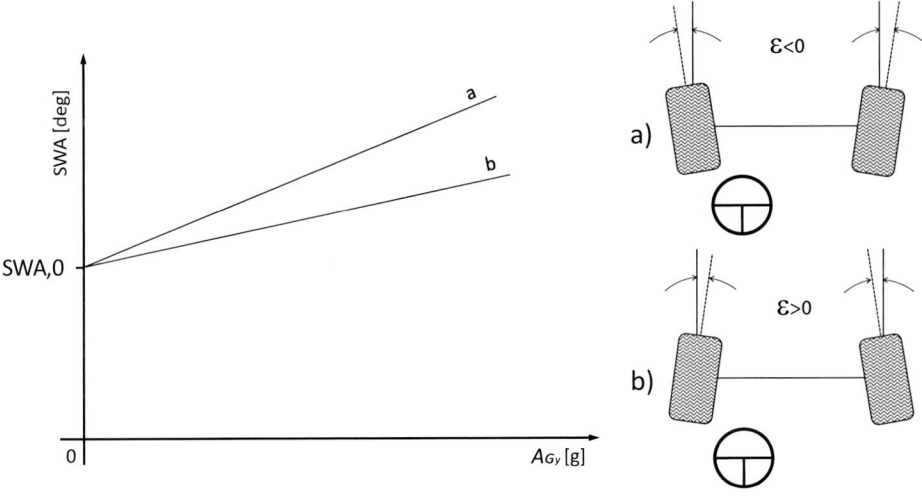

Figure 11 – Example of variation of the understeer curve as the front static *toe* varies. Curve b, more direct, is that corresponding to the front axle with *toe-in*.

On the contrary, the *toe-in* on the rear axle, in addition to increasing the slope of the understeer curve, also lowers the slope of the sideslip curve. Keeping in mind the observations just made and the tendency to *toe-out* in case of braking and to *toe-in* in case of traction, as graphically described in the following figure, we can say that the static *toe* set-up is generally oriented towards negative values on the front wheels (when they are driving) and positive on the rear wheels. On some road sports cars there is a tendency to adopt front *toe-in* (closed wheels) because, as we will see later, in this way the reactivity of the steering is lowered. Obviously, everything must be compatible with tire wear. In fact, wheels that are too open tend to wear on the inner the inner side of the tread and wheels that are too closed tend to wear on the outer side of the tread. For the latter reason, closed set-ups are used on cars with a high *Ackermann* percentage to avoid excessive internal wear of the tires. This is because they have the wheel inside the curve that tends to steer a lot, as we will see in the next chapter.

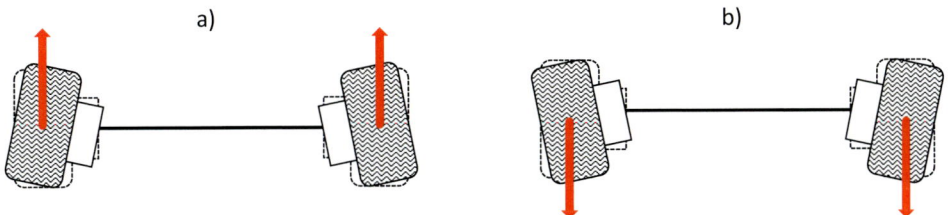

Figure 12 – Effect of traction a) and resistance to advance b) on the *toe* variation.

In support of what was stated for the responsiveness of the steering, we highlight the differences between a *toe-out* and a *toe-in* front set-up. We shall now consider the *toe-*

out set-up in the following figure in which two transverse ground forces emerge on the tires directed towards the outside (because they tend to curve in that direction) and two longitudinal forces, called *drag forces,* always directed towards the rear. The *drag forces* are the resisting forces resulting from the presence of a *toe* angle, regardless of the sign of the latter. Now we shall imagine turning the steering wheel to the left.

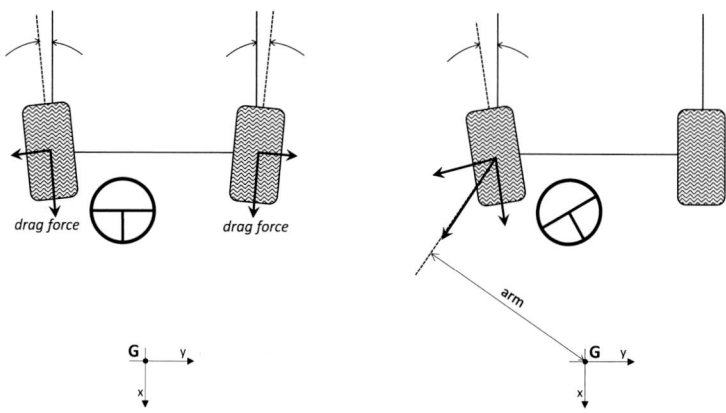

Figure 13 – Front *toe-out* set-up.

The right wheel straightens and cancels both the lateral force and the *drag force* (the longitudinal resistance to advancement remains, which we are neglecting) while the left wheel increases them. The sum of the two forces acting on the left wheel generates a yaw moment concordant with the rotation of the steering wheel. The yawing moment shown in the following figure, which for similar reasons is generated in a vehicle with a front *toe-in* set-up, is decidedly lower because it has a smaller arm and, in some situations (advanced center of gravity and very pronounced drag force), can change sign. Furthermore, considering the *drag force* effect only, we can say that the latter is certainly yawing, in the case of *toe-out* set-up, and anti-yawing in the case of *toe-in* set-up.

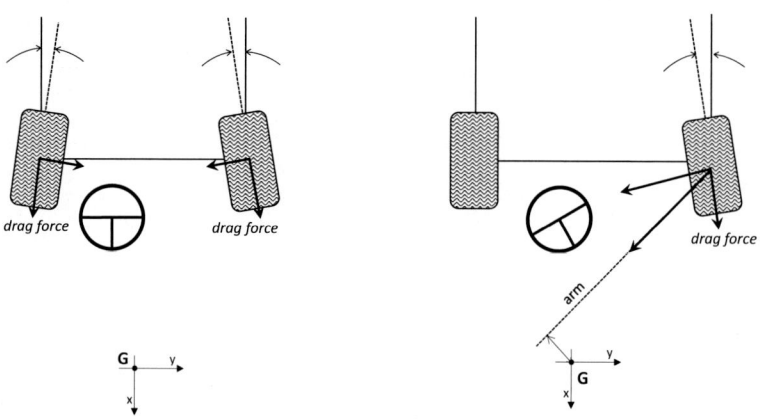

Figure 14 – Front *toe-in* set-up.

We can conclude that a front *toe-out* set-up, while increasing the stabilized understeer (with the same lateral acceleration, the corresponding steering wheel angle increases in the understeer curve), helps in those direction changes where the vehicle does not complete roll, making the steering very responsive. This happens because, having not yet formalized the load transfer, the two wheels still have the same weight in the direction of the car (see paragraph 6 of chapter 2). On the other hand, a front *toe-in* set-up is slower in changes of direction but is necessary in cars where excessive steering responsiveness is not desired. For example, in fast touring cars whose users, who for the most part some of the cases are not very expert, judge the steering readiness to be an annoying feature in straight-line driving.

We shall now develop some considerations concerning the *toe* variation in bump travel, also called bump steer (we will call rebound steer the variation in rebound travel), and the *toe* variation under lateral load. In order to stabilize the vehicle when cornering and have a concordant $\Delta\phi$ it is necessary that both in bump travel and under lateral load the front *toe* variation is negative (divergence) while the rear one is positive (convergence). In contrast with what happens with the *camber*, the $\Delta\phi$ does not depend on the difference between the performances of the axles but each axle, where possible, must contribute to giving a $\Delta\phi$ concordant with the initial steering angle. Below are the desired trends in which it is clear that the front axle opens both in bump travel and under load, and the rear axle closes in the same conditions.

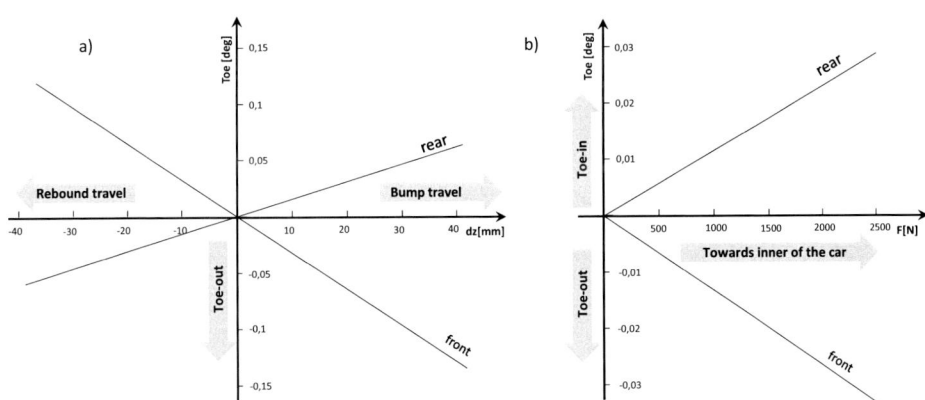

Figure 15 – Trend of *toe* variation in parallel wheel travel a) and under lateral load b).

There are rear suspensions, such as the twist beam axle, which, contrary to what was sought, open (*toe-out*) under lateral load. In this case, in order to have a concordant $\Delta\phi$, it is necessary that the front suspension opens more than the rear. The following figure shows the required trend with the evidence on the $\Delta\phi$ *(toe)* proportional to the *toe* difference that will be added to the $\Delta\phi$ due to the *camber*.

Automotive suspension

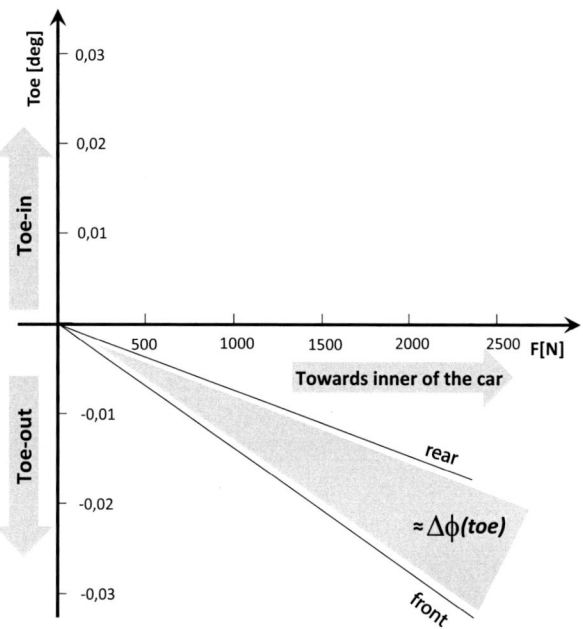

Figure 16 – Ideal *toe* trend under lateral load for a car with a rear twist beam axle.

Another effect of the wheel opening (*toe-out*) under lateral load is the increase in the stabilization time in the lateral leaning which we will discuss in detail in chapter 8. This certainly represents a defect for racing vehicles that need a fast cornering stabilization and which, for this reason, have their suspensions anchored with ball joints and not with elastic bushings. In addition, on sports vehicles, to speed up cornering, the front opening also tends to be lowered in bump travel and under lateral load. The parameter that targets the cornering speed is the delay between lateral acceleration and yaw rate, which will be explored in the study of the step steer (chapter 8). The twist beam axle is a simple and economical suspension but since it presupposes, for the reasons explained above, a great variability under load of the front *toe*, it causes the vehicle to have a significant delay in the change of direction. How this happens will become clearer after reading chapter 8.

Another very important consideration is that the trend of the *toe* variation under lateral load such as that of figure 15 has a stabilizing action on the vehicle with regard to the external actions. In fact, if a yaw speed r arises due to an external action, the vehicle would tend to turn in the direction of the latter causing a lateral force F to emerge, on the wheel outside the curve, which would tend to open the front wheel as shown in the following figure. The *toe-out* of the wheel tends to cause the vehicle to turn in the opposite direction, putting it back on track. This behavior manifests itself as a yaw resistance of the vehicle, and can therefore be classified as a stiffness to such motion.

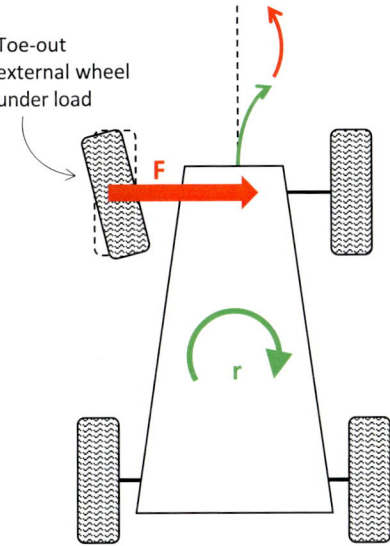

Figure 17 – Stabilizing capacity of the front wheel opening.

The same resistance occurs on the rear axle, as shown in the following figure, in which the closing of the rear wheel responds to the yaw disturbance.

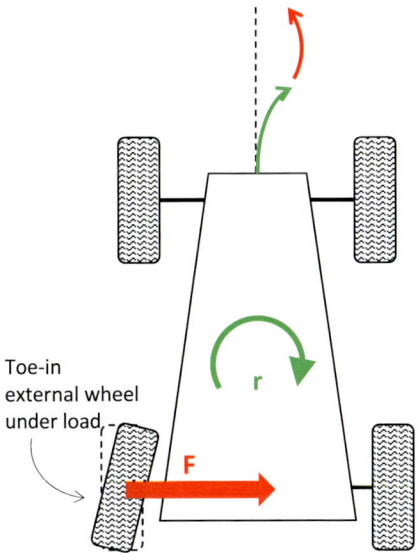

Figure 18 – Stabilizing capacity of the rear wheel closing.

We can therefore say that the front *toe-out* and the rear *toe-in* under lateral load contribute to increasing the yaw stiffness of the car. With the same steering wheel angle imposed, these last two characteristics lower the yaw speed in the stabilized phase and give greater directional stability to the vehicle in correspondence to a disturbance. The importance of these features will be even clearer when studying the step steer. In fact, in the step steer maneuver, the increase in the front opening and even more in the rear closure, produces an improvement in the yaw response. The front opening, which is important for directional stability and necessary in order to have a concordant $\Delta\phi$ in some situations, is a source of delay and it must be lowered where more sporty behavior is sought. However, all the variations under load, which are inevitable in the presence of elastic bushings, represent always delays, even if oriented towards favorable behaviors. For this reason, they are eliminated through the adoption of ball joints and very rigid bodies, where super sporty behavior is required.

6.5. The effect of the *toe* on the lateral characteristic of the tire

The variation of the *toe* when cornering, due to the lateral load and to the *bump steer*, varies the position of the longitudinal plane of the tire with respect to the longitudinal plane of the vehicle. The trace of the longitudinal plane of the vehicle on the horizontal plane is represented by the straight line *a* in the following figure. In the figure it can be seen that, due to the closure under lateral load and in bump travel, the angle α_a of the velocity V_c of the wheel center with respect to the longitudinal plane of the vehicle is less than the slip angle. The latter, which we will call α_t, is evaluated with respect to the longitudinal plane of the tire, the trace of which is the *tire center line*.

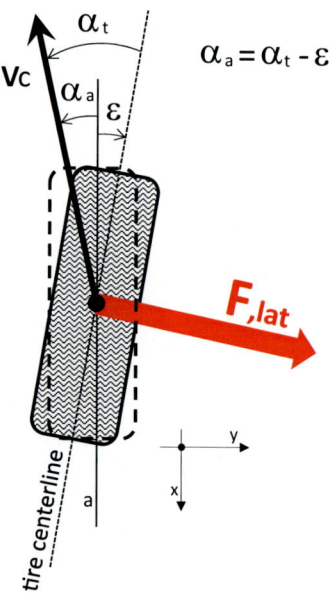

Figure 19 – Plan view of the left wheel outside the curve that closes both for lateral load and for bump travel. The displacement of the longitudinal plane of the tire towards the inside of the vehicle lowers the slip angle with respect to the axle defined with respect to the straight line a.

The angle α_a represents the slip angle of the axle and is the quantity that defines the lateral behavior of the vehicle. This is because it measures the actual lateral displacement of the wheel center with respect to the longitudinal plane of the vehicle. For this reason, the goal is to trace the characteristic of the tire with respect to α_a, which we will call *characteristic with respect to the axle*, in order to identify the variation of the slip stiffness and the position of the new *sliding limit* as a function of the variation of *toe*.
Being:

$$\alpha_a = \alpha_t - \varepsilon$$

to obtain the characteristic curve with respect to the axle, we shall just subtract the *toe* variation ε from the slip angles of the tire characteristic, for each lateral force value. The result is the bold curve shown in the following figure which, in this case elastokinematic (*toe-in* under lateral load and in bump travel), is more rigid because, with the same lateral force, the slip angle with respect to the vehicle is always lower.

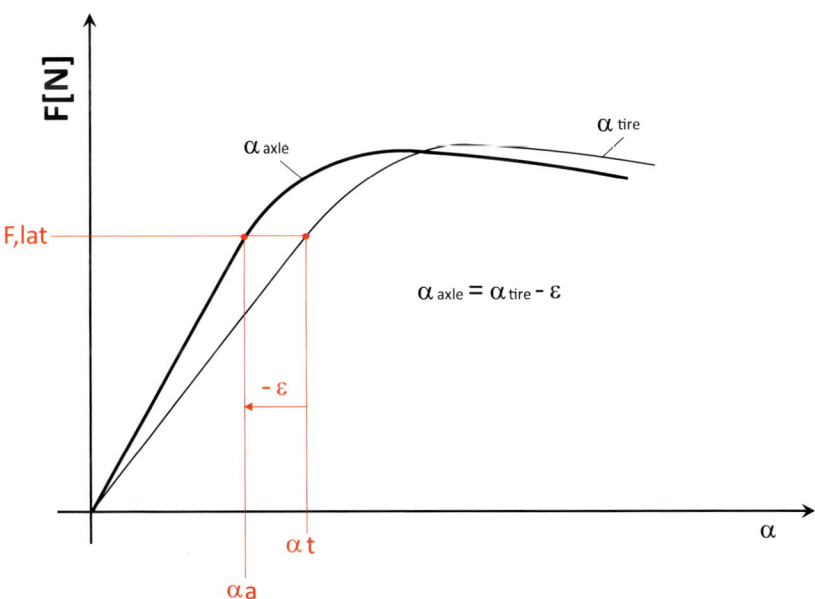

Figure 20 – Transformation of the characteristic of the tire into the characteristic with respect to the axle in case of wheel closure under lateral load. Qualitative representation.

Note that the *toe* variation depends both on the lateral force and on the *bump steer*. At low slip angles the amount to be subtracted, to obtain the characteristic with respect to the axle, tends to cancel out. Similarly, the characteristic with respect to the axle can be evaluated in the case of *toe-out*. In the following figure, the characteristics with respect to the axle in case of closing under load and bump travel (*Toe-in, var*), and in case of opening (*Toe-out, var*) as function of α_a are shown.

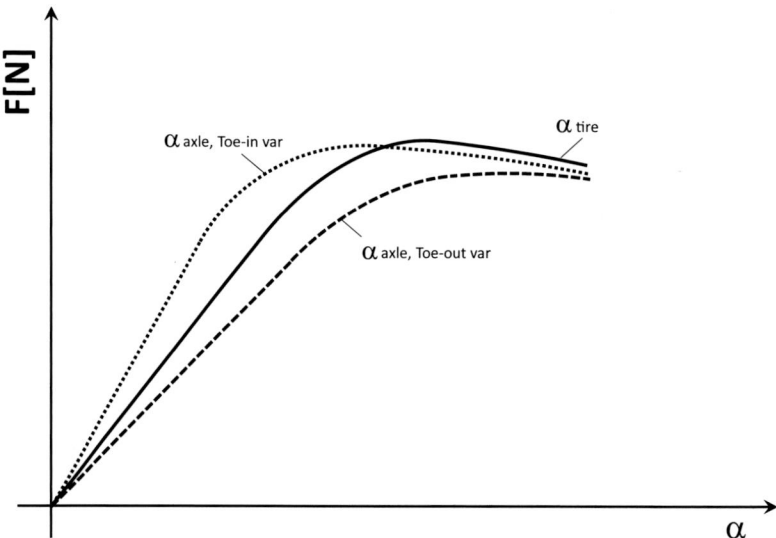

Figure 21 – Qualitative representation of the tire characteristic with respect to the axle in case of *Toe-in*, var and *Toe-out*, var.

For the static convergence only, the curves translate leftwards in the case of *Toe-in* and rightwards in the case of *Toe-out*, as shown in the following figure. With the same lateral force the static *Toe-in* produces a smaller slip angle while static *Toe-out* produces a greater slip angle. We will detail this last statement below.

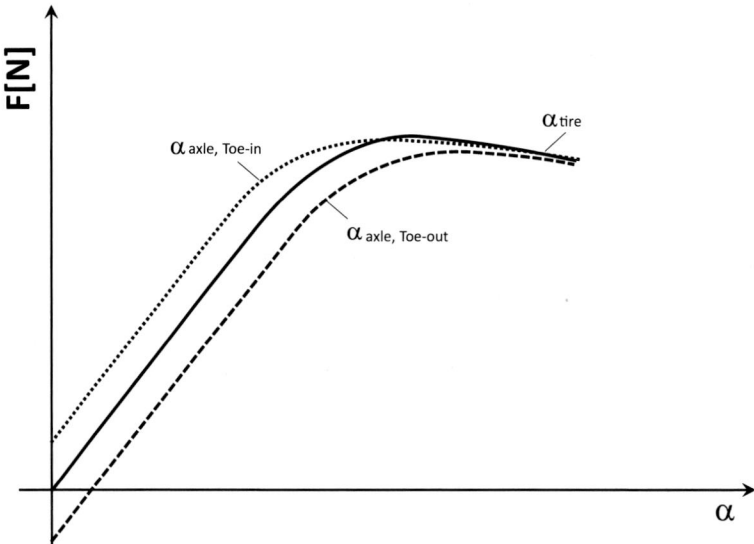

Figure 22 – Qualitative representation of the characteristic of the tire with respect to the axle in case of static *Toe-in* and *Toe-out*.

Suspension – Set-up and operation

In the following figure (on side a) a left front tire with positive *toe* angle (closed wheel) is shown. Since the vehicle travels straight, the speed direction of the tire is parallel to the straight line *a*, and is inclined by the angle ε with respect to its longitudinal plane. For this reason, a lateral force emerges perpendicular to the centerline of the tire and directed towards the inner of the vehicle, which is equal to the positive force corresponding to the zero α_a of the previous curve relating to the *Toe-in*.
In fact, it turns out:

$$\alpha_t - \varepsilon = 0$$

and then:

$$\alpha_t = \varepsilon$$

The other side of the figure (side b) shows a left front wheel with negative (open) *toe* angle in which the transverse force corresponding to zero α_a is negative because it is directed towards the outside of the vehicle. This is because it is always perpendicular to the center line of the tire and belongs to the half plane which does not contain the velocity vector.

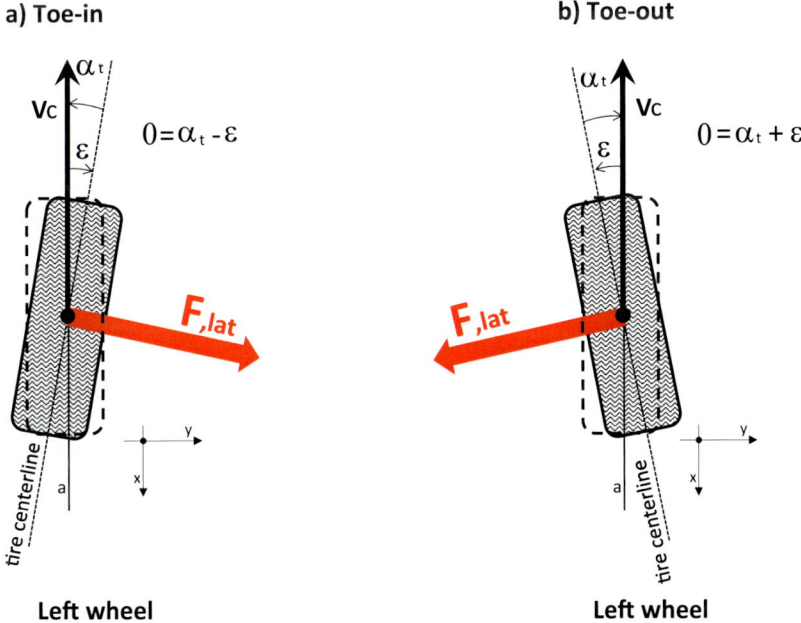

Figure 23 – a) Plan view of the left wheel (external to a possible right cornering) of a car driving straight with static *Toe-in*. The slip angle with respect to the axle is zero and the emerging transverse force is directed towards the inside of the vehicle. b) Plan view of the left wheel (external to a possible right cornering) with *static Toe-out*. The slip angle with respect to the axle is zero and the transverse force is directed towards the outside of the vehicle.

For these reasons, the characteristic of the tire with respect to the axle is shifted leftwards in the case of static *Toe-in* and rightwards in the case of static *Toe-out*.

Automotive suspension

6.6. *Toe* variation under longitudinal load

The change in *toe* under longitudinal load is important for the control of the vehicle during braking and acceleration maneuvers. The braking maneuver is simulated with the application of the force at the center of contact patch, while the obstacle passage with the force applied to the center of the wheel, both directed towards the rear. On the other hand, traction is simulated with the application of the force to the center of the wheel in vehicles equipped with drive shafts and on the ground in the presence of chain transmission, both directed according to the direction of travel. The latter case refers to go-karts and many self-built vehicles with motorcycle engines. As for braking, the ideal condition for a front suspension is that in which the wheels open (*toe-out*) because, if the maneuver is carried out in a curve, the opening of the outside wheel tends to increase the radius of curvature by increasing the understeer. In the event of a converging external wheel, the vehicle would oversteer and risk spinning. Note that the increase in understeer is the most desired condition. This is because the increase in the steering wheel angle, which is the emergency maneuver that the user must perform, is perfectly aligned with what they would instinctively do. In the case of oversteer, the user should perform counter-steering which is not an instinctive maneuver for the less experienced. For similar reasons, closing under braking is the ideal condition for rear suspension.

The arrangement of the kinematic mechanism which allows divergence during braking is that in which the steering rod *b* is inclined, as in the following figure, with respect to the straight line *a* which joins the center of the ball joint 1 of the arm and the center of the bushing 11 anchored to the frame. In this configuration, the steering rod rotates around point 22 by moving point 2 towards the inside of the vehicle so that the wheel opens.

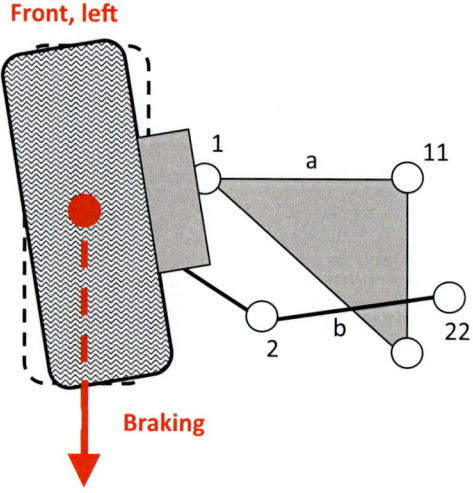

Figure 24 – Plan view of a divergent suspension under braking. The dotted figure represents the wheel with no load applied.

Another way to obtain the divergence when braking is to lower the stiffness in the transverse direction of the bushing 22. This can be achieved by eliminating rubber in that direction but paying attention not to couple the frequency of the *toe* variation motion under braking with that of the *toe* variation motion under lateral load.

Ideally, the front suspension should have *toe-out* even in traction (acceleration) because the outer wheel would help to increase the radius of curvature in case of application of torque when cornering. Since it is difficult to have both the performances, *toe-out* is preferred when braking, because braking is an emergency maneuver and it is more difficult to manage. In any case, braking and traction in cornering lower the adherence capital in the transverse direction of the tire, and also tend to increase the slip angle and therefore the understeer.

Whatever the performance that can be achieved, it is very important that the behavior of the convergence both in braking and in traction (acceleration) is perfectly symmetrical. This is because small differences between left and right could cause significant lateral displacements of the car under braking and drift under acceleration, as shown in chapter 9.

6.7. The kinematics of the suspension

The suspension consists of a kinematic mechanism that defines the positions assumed by the wheel with respect to the vehicle body during the permitted motions. It is composed of the upright, the arms and the related connection constraints. The relative positioning between the arms and the type of constraints defines the motion of the wheels with respect to the vehicle body which, until the elastic bushings are introduced, will be considered a rigid motion.

Like all acts of rigid motion, the movement of the wheel with respect to the vehicle body is comparable to a translation and a rotation with respect to an axis of instantaneous rotation, variable over time. The understanding of this last statement presupposes the knowledge of the kinematics of the rigid motions, but it is not difficult to imagine that the kinematics imposes a rotation on an axis and not a vertical translation, given the variation of inclination of the wheel during the movement relative to the vehicle body. The position of this axis, as we will see in detail below, depends on the relative position of the arms, on the type of constraints and is variable as the suspension configuration varies. The following figure shows the right front wheel sectioned by a longitudinal plane and a plane transversal to the vehicle passing through the wheel center. The axis of instantaneous rotation intersects the transverse plane and the longitudinal plane respectively in the *Instant center front view* (ICFV) and in the *Instant center side view* (ICSV)[1], which we will call the suspension pivots.

---

[1]. The acronyms ICFV and ICSV were taken from the book W.F. Milliken, D.L. Milliken, *Race Car Vehicle Dynamics*, SAE, 1994.

Automotive suspension

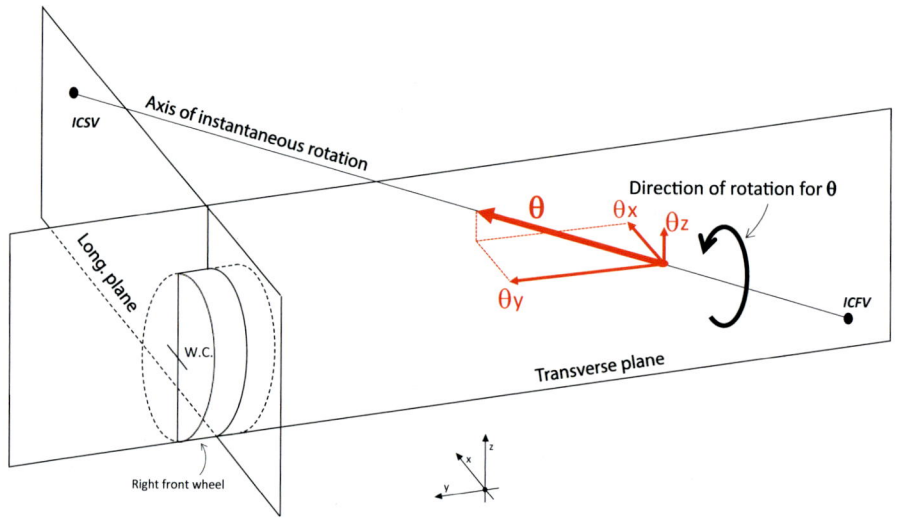

Figure 25 – Schematization of the axis of instantaneous rotation of the right front wheel.

From what has been said, we can schematize the movement of the wheel with the rotation vector θ parallel to the axis of instantaneous rotation and directed towards the *ICSV*. The rotation vector θ, anticlockwise looking from the ICSV towards the ICFV, can be decomposed into the three rotations θx, θy and θz along the three cartesian axes represented in the figure. The rotation θx around the x axis reproduces the variation of *camber*, the rotation θz around z reproduces the variation of *toe* and the rotation θy around y reproduces the variation of *spin*. To clearly understand these last statements, it is necessary to assume that the rotation around an axis is a vector parallel to the axis itself and it is positive when it is counterclockwise looking at it from the tip of the vector towards the base. Depending on how the axis of Instantaneous rotation is inclined, the values of the three projections θx, θy and θz of the vector θ on the three axes change. In the following figure the right front wheel with the three rotations associated with the variations of the characteristic angles is shown.

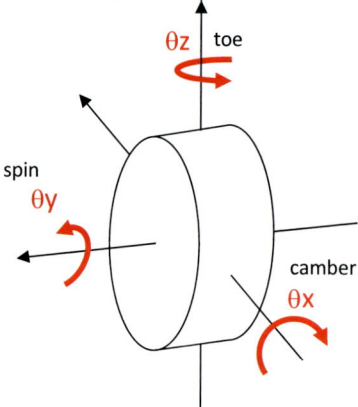

Figure 26 – Projections on the three axes of theta rotation around the axis of instantaneous rotation.

Suspension – Set-up and operation

The inclination and position of the axis of instantaneous rotation, which defines the kinematic behavior of the suspension, depends on the location of the *ICFV* and *ICSV*. It varies as the position assumed by the wheel with respect to the vehicle body changes. We shall now describe the operational method for their identification and regulation.

The *ICFV* is the instantaneous rotation center of the plane kinematics obtained by projecting the suspension kinematics on the transverse plane to the vehicle. The kinematics projected on the transverse plane for a McPherson suspension looks like the following figure.

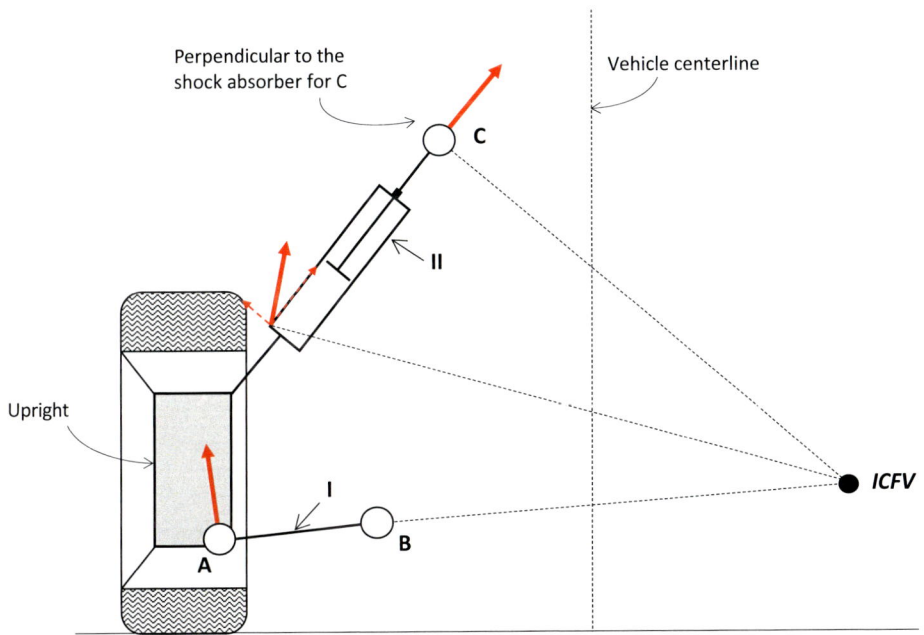

Figure 27 – Construction of the ICFV for a McPherson.

In this architecture, the set consisting of wheel, upright and shock absorber (only the strut housing without the piston rod) is considered a single part with the vehicle body constrained by the link I, the translational joint II and the ball joint in C. The intersection of the perpendiculars to the displacements imposed by these constraints, which are respectively the extension of link I and the perpendicular to the shock absorber for C, is the instantaneous rotation center of the aforementioned assembly. The understanding of this requires knowledge of the planar rigid motions which are always comparable to a rotation around a center obtained by intersecting the perpendiculars to the displacements of at least two points of the part (*wheel-upright-shock absorber*). In this case, the link I imposes at the point *A* of the upright a movement perpendicular to its median, while the combination between the translational joint and the ball joint on the piston rod of shock absorber imposes at point *C* a displacement parallel to the translational joint. To understand this last statement, it is necessary to consider that point *C* belongs to the

145

strut housing of the shock absorber and, since it lies on the center of ball joint, does not undergo any lateral displacement induced by rotation. Otherwise, any other point of the shock absorber also has a perpendicular displacement to the translational joint due to the rotation around C, as shown in the figure.

The analogous construction for the Double wishbone is illustrated in the following figure, in which the extension of link II replaces the perpendicular to the shock absorber.

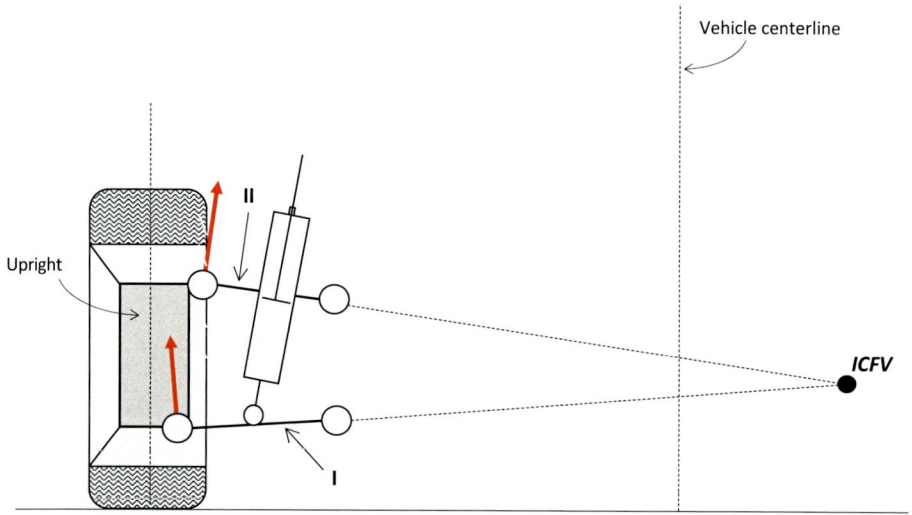

Figure 28 – Construction of ICFV in Double wishbone suspension.

To identify the links defined above it is necessary to reduce the control arms to two unique equivalent links and project them on the transverse plane. The following figure illustrates the construction that consists in replacing the arm identified by the three joint centers *11*, *22* and *55* with the reduced link *55-55** perpendicular to *11-22* and with a pin in *55**.

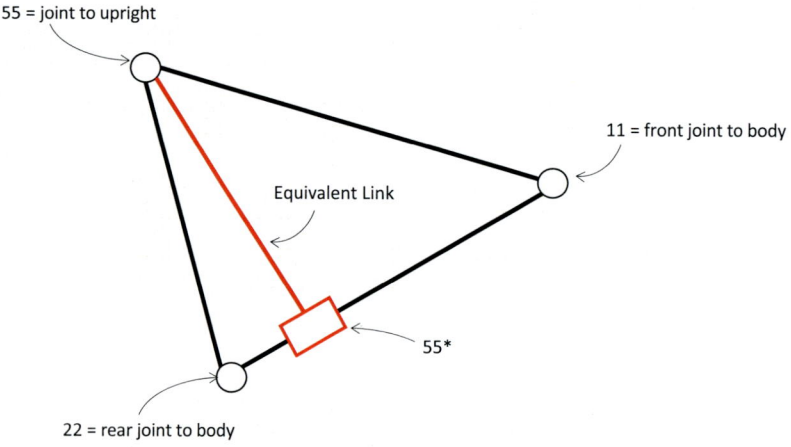

Figure 29 – Construction of equivalent link of control arm.

Suspension – Set-up and operation

We shall now shift to the *ICSV* definition. This is the center rotation of the suspension kinematics, projected on the longitudinal plane of the vehicle. As shown in the front McPherson scheme in the following figure, the *ICSV* is the intersection between the extension of the link *A*, built as described above, and the perpendicular to the projection of shock absorber (projection of translational joint) passing through *C*. In identifying the *ICSV*, link *A* of the control arm can be replaced by the line parallel to the projections of *11* and *22* (previous figure) on the longitudinal plane, passing through the projection of point *55* on the same plane. This construction helps us when the equivalent link, previously seen, turns out to be almost perpendicular to the longitudinal plane and consequently the two projections are unable to identify a well-defined segment to be extended.

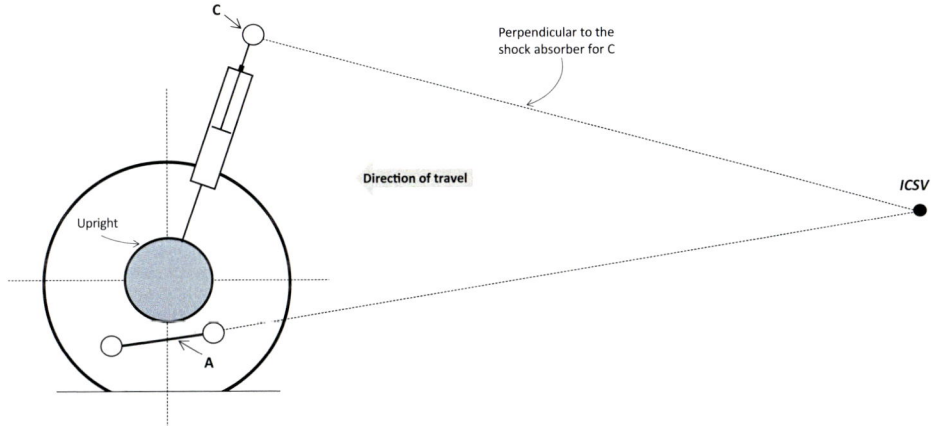

Figure 30 – Construction of the ICSV for a front McPherson.

Similar construction for the Double wishbone in the following figure

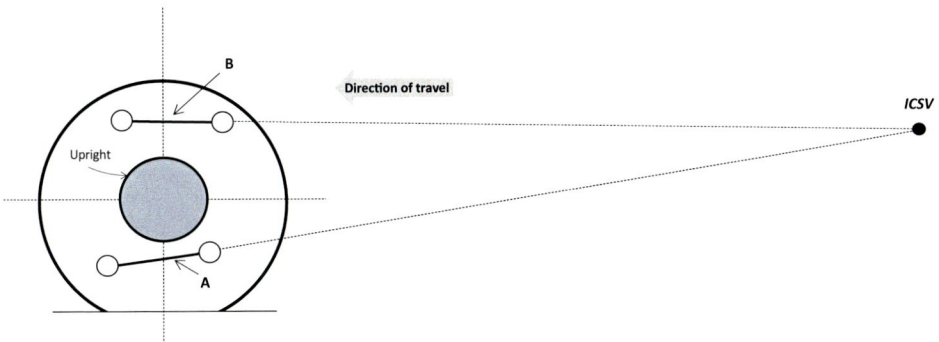

Figure 31 – Construction of the ICSV for a front Double wishbone.

The graphical constructions seen show that the positioning of the two suspension pivots, which defines the position and orientation of the axis of instantaneous rotation, depends on the positions and inclinations of the arms and telescopic shock absorbers.

We shall then analyze the case of a counterclockwise rotation (looking at the *ICFV* from the *ICSV*) of the vector θ of Figure 25 which tends to lift the right front wheel upwards. We observe that the *ICSV* is higher than the *ICFV*. Consequently the vector θz, which represents the rotation of the wheel around z and therefore the *toe* variation in bump travel, is positive and facing upwards. This means that the tire, seen from above, rotates around z counterclockwise and then closes (*toe-in*), since we are talking about the right front wheel. To obtain the desired opening of the front wheel in bump travel it is necessary to vary the inclination of the arms so that the *ICSV* is lower than the *ICFV*. Another characteristic of the kinematics in the figure 25 is that θx is oriented towards the rear of the car, and this means that we have *camber* gain in bump travel because, looking at the car from behind, the rotation of the tire is counterclockwise.

A very effective technique in setting a suspension is to identify the two pivots according to the aforementioned specifications, then trace the axis of Instantaneous rotation and draw the planes of the arms (A, B in the following figure) and the lines of the steering rod so that all converge on the instantaneous axis. The identification of the position of the two pivots presupposes the following considerations:

- With reference to Figure 25, it shall be noted that to move the wheels upwards it is necessary to perform a counterclockwise rotation on the right side and a clockwise rotation on the left side, looking at the corresponding axis of instantaneous rotation from the *ICSV* to the *ICFV*. Consequently, the components θx and θz assume opposite signs between the left side and the right side. For example, *camber* gain on the right means positive θx (directed towards the rear) while *camber* gain on the left means negative θx.
- To increase the *camber* gain in bump travel *(Dcamber/dz)* the θx component of the vector θ must be increased. This is achieved by decreasing the *y* of the *ICFV* or increasing the *x* of the *ICSV* compatibly with the *anti dive, anti lift* or *anti squat* needs that will be treated later. The most sensitive factor to the variation of *camber* is the y of the *ICFV* as will be amply demonstrated below.
- In the front suspensions, where the *toe-out* (wheel opening) in bump travel is required, the θz component must be positive on the left and negative on the right (*ICSV* lower than the *ICFV*), while in the rear suspensions must be the opposite. Each of these needs is met only when, on each axle, the *ICSV* is lower than the *ICFV*. This last statement derives from the need, due to the longitudinal kinematics constraints that will be described in the last paragraphs, to position the front *ICSV* set back with respect to the front wheel center and the rear *ICSV* advanced with respect to the rear wheel center. If it is not possible to position di *ICSV* below the respective *ICFV*, the opening or closing of the wheel in bump travel can also be adjusted by changing the z of one of the two ball joints of the steering rod (or of another joint in the case in which the suspension is not a Double wishbone or a McPherson). In this case, however, the *toe* variation in bump travel tends to be less linear. Alternatively, by adjusting the heights of the two pivots and passing the median of the steering rod on the instantaneous axis, more linear trends are obtained.

Suspension – Set-up and operation

The following figure shows, planes A and B whose intersection identifies the axis of Instantaneous rotation for two suspension architectures. In the Double wishbone the planes A and B are the planes of the arms while in the McPherson plane B is the plane perpendicular to the axis of the shock absorber passing through the center of the ball joint in the vehicle body. Obviously in both cases the median straight line of the steering rod passes through the intersection between the planes.

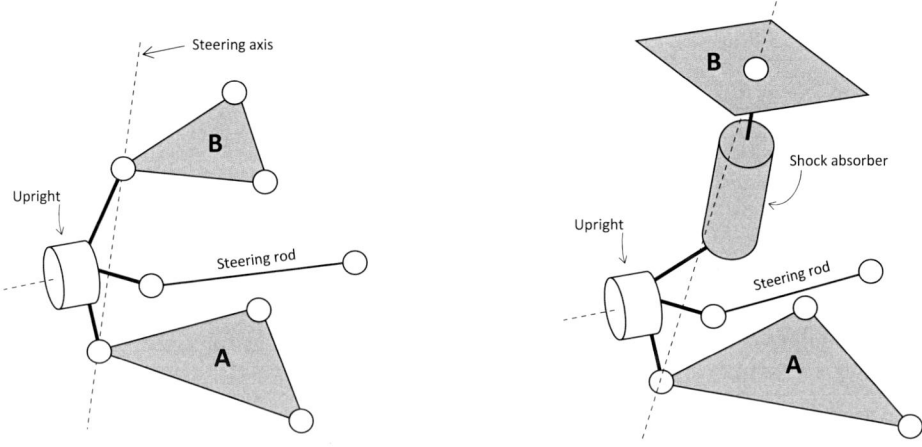

Figure 32 – Planes A and B which identify the axis of Instantaneous rotation for McPherson and Double wishbone.

6.7.1. The roll center

The roll center of an axle is the intersection point between the roll axis and the transverse plane of the vehicle passing through the wheel center. It is located along the center line of the vehicle. Its height from the ground is a quantity to be kept under control because it guarantees a load transfer quota that manifests itself instantaneously unlike the quota due to the roll that undergoes the delay due to the work of the springs and shock absorbers, as fully described in chapter 3. In order to reduce the difference between the two aforementioned quotas, which, due to the different intervention times, would cause a discontinuous behavior of the load transfer, the height of the roll center must be correlated to the height of the center of gravity. The lower the center of gravity of the vehicle, the lower the roll center must be. In production cars, with center of gravity around 600mm from the ground, the front roll center varies from 40 to 90 mm and the rear is a consequence of the fact that the roll axis must have an inclination ranging between 1 and 1.5 deg. A higher roll axis would accelerate the load transfer making the car more responsive but, at the same time, increases the contribution of *suspension jacking* and decreases the weight of roll stiffnesses distribution in the distribution of load axle between the axles.

As already mentioned, the rear roll center must be higher than the front roll center since a roll axis tilted forward, in addition to being closer to the main axis of inertia, favors cornering, by generating a gyroscopic torque of yaw concordant to the curve. The ideal condition is the coincidence of the vehicle roll axis with the main axis of inertia, to avoid coupling between the roll and pitch motion. Unfortunately, this is not easy because the position of the main axis is difficult to identify. To avoid such couplings, it is necessary to ensure that the variation in the y and z directions of the front roll center is similar to the rear one so as not to generate excessive variations in the inclination of the roll axis. The reasons for these statements can be explained by referring to the rotor in the following figure which rotates around the axis a of its bearings.

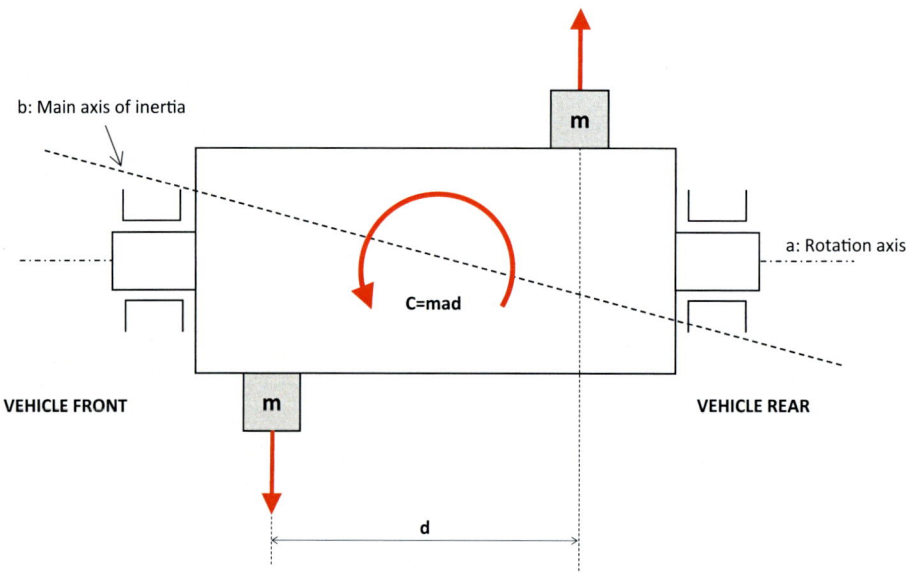

Figure 33 – Schematic representation of the vehicle with a rotor rotating around the roll axis.

If we apply the two masses m to the rotor, we displace its main axis of inertia b which initially coincided with a. If we imagine that the vehicle is the rotor and the axis of rotation a is the roll axis, we can say that the roll axis a is more inclined towards the front axle than the main axis of inertia, and a rotation around a generates an unbalanced pitch torque which loads the front axle and lightens the rear axle. With this example, it is shown that the variation in the inclination of the roll axis generates a coupling between the roll motion and the pitch motion. If we imagine rotating the main axis of inertia around the z axis, we simulate the lateral movement of the roll axis, which in this case generates the coupling between the roll motion and the yaw motion.

Now we shall calculate the position of the roll center, that is the instantaneous center of rotation of the vehicle body in a transverse plane passing through the wheel center. The mathematical method for identifying it presupposes considerations concerning rigid relative plane motions. These are not discussed here to avoid making the description too complex and assuming the reader is already familiar with these concepts. From a practical point of view, the roll center is obtained graphically by intersecting the connecting point between the point Ct (center of the contact patch) and the *ICFV* with the center line

Suspension – Set-up and operation

of the car, as shown in the following figure. Knowing the angle φ we can write:

$$hRc = \frac{t}{2} \tan(\varphi)$$

where t is the wheel track at ground, Ct is the center of the contact patch and φ follows the relation:

$$\tan(\varphi) = \frac{ICFV,z - Ct,z}{ICFV,y - Ct,y}$$

but it can also be calculated by measuring the variation of wheel track dy at ground as a function of bump travel dz according to the following formula:

$$\tan(\varphi) = \frac{dy}{dz}$$

In the following figure, the construction of the displacement s of the point Ct due to the rotation Δy around the $ICFV$ induced by the bump travel of the wheel.

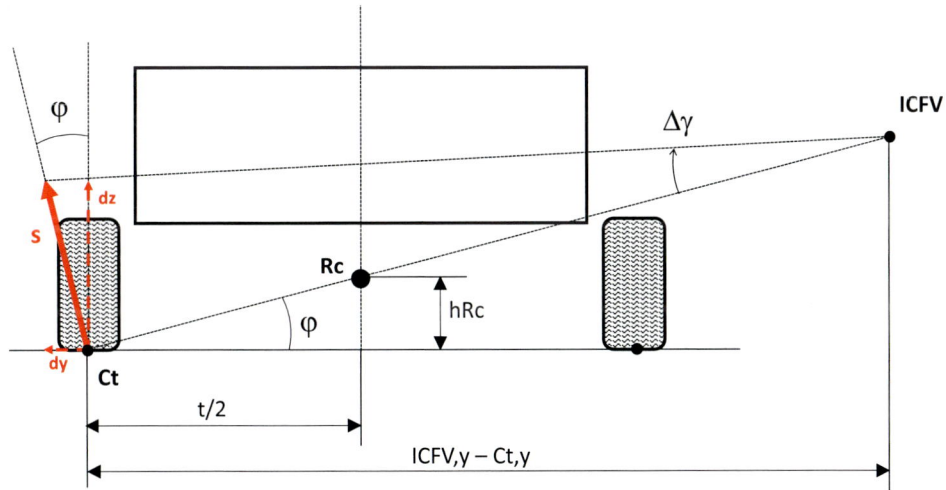

Figure 34 – Graphic construction of the roll center and the displacement of the center of the contact patch.

This rule applies to any suspension architecture, even when the *ICFV* is improper. In this latter condition, the *Rc* is identified by intersecting a straight line parallel to the improper center and passing through *Ct* with the center line. This circumstance occurs, for example, when a Double wishbone has parallel arms and the *ICFV* is the improper center (the common direction) of the two parallel lines that are obtained by intersecting the planes of the arms with the transverse plane to the vehicle. If the arms of the Double wishbone are parallel to the horizontal plane, the line *Ct-Rc* is parallel to the ground line and the *Rc* has zero height.

6.7.2. *Camber* variation in bump travel

The desired *camber* variation in bump travel is a consequence of the *ICFV* position. With the help of the following figure it is understandable that, once the roll center height is defined, by approaching the *ICFV* closer to the roll center along line *a*, greater *camber* gain is obtained with the same vertical displacement of *Ct*. In fact, since the wheel motion with respect the vehicle body is a rotation around the *ICFV*, the displacement *s* of the contact point at ground is:

$$s = \Delta \gamma D(ICFV - Ct)$$

In which the rotation $\Delta\gamma$ of the rigid planar motion precisely represents the *camber* variation sought, while *D* represents the distance between the points contained in parentheses. Applying the minus sign to the $\Delta\gamma$, because it is a *camber* gain, the vertical displacement is:

$$dz = s \cos \varphi = - \Delta\gamma D(ICFV - Ct) \cos \varphi$$
$$D(ICFV - Ct) \cos \varphi = (ICFV,y - Ct,y)$$
$$dz = - \Delta\gamma (ICFV,y - Ct,y)$$

and then:

$$\frac{\Delta\gamma}{dz} = - \frac{1}{(ICFV,y - Ct,y)}$$

From which it is clear that the *camber* gain in bump travel increases when we bring the *ICFV* closer to the wheel. On the other hand, when the *ICFV* lies to the left of the wheel (outside to vehicle wheel track) there is a loss of *camber*. If the origin of the reference system is in Ct we can write:

$$\frac{\Delta\gamma}{dz} = - \frac{1}{ICFV,y}$$

Which is a more convenient formulation for the developments to follow.

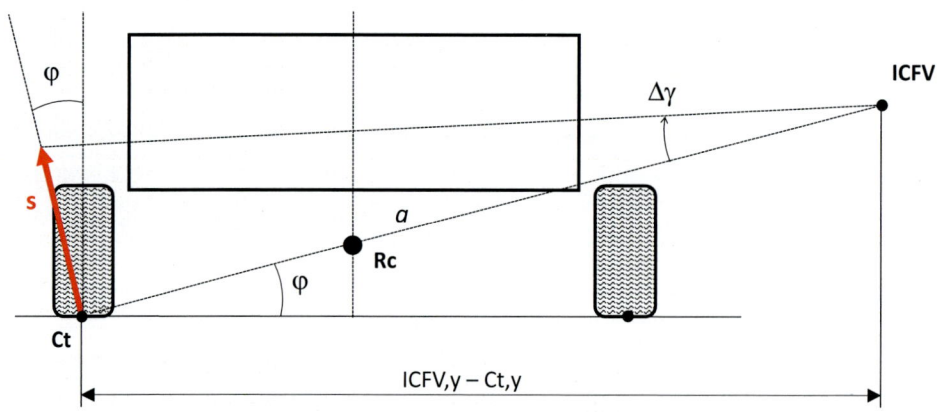

Figure 35 – Graphic construction of *camber* gain.

6.7.3. Camber variation when cornering

In the previous paragraphs we have stated that, in the stationary condition, the distribution of the slip stiffness between the axles depends on the difference between the *camber* gains in bump travel. In reality, the roll of the vehicle when cornering, even if it causes the external wheels to bump and the internal ones to rebound, in most cases produces a loss of *camber* on the outside wheels. This is because the roll of the vehicle body, in addition to bump of external wheel and the rebound of internal one, gives also rise to rotations around the respective centers of contact patches. To reproduce the kinematics of the wheels during the roll, the bump and rebound motions are not enough but it is necessary to add other considerations.

Below is a description of the roll kinematics of the vehicle, which aims to identify the factors on which *camber* loss depends. The figure below shows the view, from the rear of the vehicle, of an axle, which rolls by an angle Ψ, in which the kinematic chain of the left suspension is only shown in order not to complicate the drawing. In the kinematic study, the *ICFV* point is the center of the relative motion between suspension and the vehicle body and, as such, is considered as a common point.

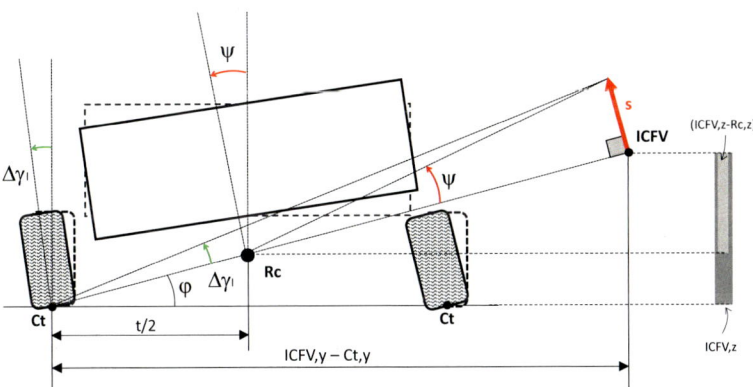

Figure 36 – Rear view of Kinematic chain of the roll motion of a car cornering to the right with evidence of the loss of *camber* of the outer wheel.

The roll produces a displacement s of the *ICFV* center which, seen as belonging to the vehicle body, derives from the rotation ψ around the roll center Rc. Since the *ICFV* center also belongs to the suspension and the displacement s is common to both the suspension and the vehicle body, it is possible to calculate the consequent rotation $\Delta\gamma_i$ of the wheel, which represents the loss of *camber* in roll. If *ICFV* belongs to the vehicle body (remembering that the D operator represents the distance between the points contained in the brackets) we can write:

$$s = \psi D(ICFV - Rc)$$

If it belongs to the suspension:

$$s = (\Delta\gamma_i) D(ICFV - Ct)$$

Automotive suspension

by imposing equality between the displacements it results:

$$\psi D(ICFV - Rc) = (\Delta \gamma_l) D(ICFV - Ct)$$

$$\Delta \gamma_l = \psi \frac{D(ICFV - Rc)}{D(ICFV - Ct)}$$

Making some proportions relative to the right triangles in the figure:

$$\frac{D(ICFV - Ct)}{ICFV,z} = \frac{D(ICFV - Rc)}{(ICFV,z - Rc,z)}$$

$$\frac{D(ICFV - Rc)}{D(ICFV - Ct)} = \frac{(ICFV,z - Rc,z)}{ICFV,z}$$

and then the *camber* loss of the left wheel:

$$\Delta \gamma_l = \psi \frac{(ICFV,z - Rc,z)}{ICFV,z}$$

which turns out to be proportional to the ratio between the height difference of the *ICFV* and the *Rc* (light gray tower) with respect to the height of the *ICFV* itself (dark gray tower). Since the modulus of this ratio is always less than one, the loss of *camber* is always less than the roll angle and it is a percentage of this latter. It is clear that approaching the *ICFV* to *Rc* along their junction decreases the loss of *camber* in roll, as shown in the following figure.

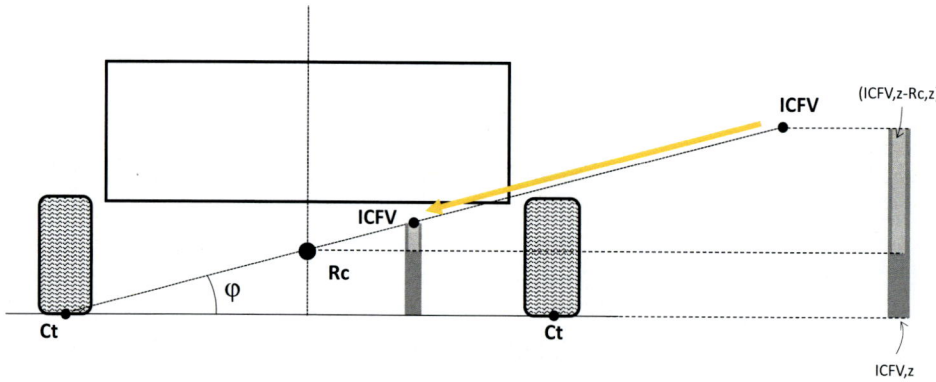

Figure 37 – Displacement of the ICFV towards the corresponding wheel. The ratio between the light gray and dark gray tower is lowered with the consequent less loss of *camber* in roll.

When the *ICFV* crosses the *Rc* (it is located between *Rc* and *Ct*) the aforementioned ratio becomes negative and *camber* gain in roll is obtained. This situation, sometimes favorable to cornering dynamics (far from the limit), becomes difficult to manage as the consequent high *camber* gain in bump travel and the large variation of wheel track cause unbalance with the other axle and annoying lateral skidding. These can lead to sudden crisis in situations close to the limit. Furthermore, it should be noted that, since the following proportion is valid within the right triangles of the previous figure, the loss of *camber* can only be modulated through the *ICFV,y* and not through the *ICFV,z*. If the *ICFV* is lowered, by varying the position of the roll center, the loss of *camber* does not change.

$$\frac{(ICFV,z - Rc,z)}{ICFV,z} = \frac{(ICFV,y - t/2)}{ICFV,y}$$

By reworking the relationship as follows, we see that it is possible to derive the loss of *camber* in roll from the formula of the gain of *camber* in bump travel.

$$\Delta\gamma_l = \psi \left(1 - \frac{t/2}{ICFV,y}\right)$$

We observe that for *ICFV,y* > *t/2* (origin in *Ct* of the outer wheel) the loss of *camber* is a fraction of the roll angle. On the other hand, when ICFV,y < t/2 the quantity in brackets becomes negative and the loss becomes gain. By replacing the expression of *ICFV,y* of the gain of *camber* in bump travel, we obtain the loss of *camber* in roll as a function of $\Delta\gamma$.

$$\Delta\gamma_l = \psi \left(1 + \frac{t}{2}\frac{\Delta\gamma}{dz}\right)$$

remembering that the *camber* gain $\Delta\gamma$ is negative while $\Delta\gamma_l$ is positive because it agrees with the roll. From the formula it follows that suspensions that do not have *camber* gain in bump travel (*ICFV* improper or *ICFV,y* = ∞) have a loss of *camber* in roll equal to the roll itself. It shall be noted that these mathematical elaborations are allowed for small roll angles, otherwise it is also necessary to consider the displacements of the *ICFV* and of the *Rc* during the evolution of the roll kinematics and of the bump travel kinematics. Furthermore, it shall be remembered that the calculations carried out do not take into account the losses of *camber* related to the loads present in the curve (lateral, longitudinal and vertical).

The next step is to appropriately balance the loss of *camber* between the front and rear axles in order to obtain a $\Delta\phi$ of steering wheel angle concordant with the set one. For this to happen, the loss of front *camber* must be greater than the rear. It must therefore turn out:

$$\Delta\gamma_l, f > \Delta\gamma_l, r$$

by replacing the expressions of the *camber* loss it results:

$$\frac{(ICFVf,z - Rcf,z)}{ICFVf,z} > \frac{(ICFVr,z - Rcr,z)}{ICFVr,z}$$

$$1 - \frac{Rcf,z}{ICFVf,z} > 1 - \frac{Rcr,z}{ICFVr,z}$$

$$\frac{R_{cr,z}}{ICFV_{r,z}} > \frac{R_{cf,z}}{ICFV_{f,z}}$$

where f and r next to the quantities indicate that they are respectively front and rear. Substituting the heights as a function of the coordinates in y

$$ICFV, z = (ICFV, y) \tan \varphi$$

$$R_{c,z} = \frac{t}{2} \tan \varphi$$

we get:

$$\frac{(t_r) \tan(\varphi,r)}{(ICFV_r,y) \tan(\varphi,r)} > \frac{(t_f) \tan(\varphi,f)}{(ICFV_f,y) \tan(\varphi,f)}$$

simplifying

$$\frac{t_r}{ICFV_r,y} > \frac{t_f}{ICFV_f,y}$$

we get the definitive expression

$$ICFV_r,y < ICFV_f,y \frac{t_r}{t_f}$$

When the ratio between the wheel tracks on the two axles is about 1, taking into account the formula of the *camber* gain that we rewrite below for convenience, we can say that a greater gain of *camber* on the rear is enough to balance the roll losses and have a $\Delta\phi$ concordant.

$$\frac{\Delta y}{dz} = -\frac{1}{ICFV,y}$$

remembering that it is always recommended to measure the loss of *camber* on each axle to verify the angles at which the tires face the curve. Contrarily, in cars with large differences between the wheel tracks it is necessary to apply the previous inequality which, for lower rear wheel tracks, will produce a further increase of *camber* gain at the rear axle.

An important result of this analysis is the confirmation that the knowledge of the distribution of *camber* gains in bump travel allows to predict the distribution of the losses in roll but not the values of the losses of each axle that define the interaction between tire and road. In the study of suspension kinematics, the *camber* gains are only a consequence of the sizing of the desired losses in roll. Generally, we start from the identification of the maximum permissible loss on the front axle at maximum lateral acceleration (as a percentage of the maximum roll angle that is known) and the consequent rear loss is defined, also taking into account the variation of *camber* under lateral load and when steering. In this regard, we reformulate the previous expression to derive the *ICFVa, y* as a function of the loss of *camber* in roll.

$$\Delta \gamma_{,,}f = \psi \left[1 - \frac{t_f}{2(ICFVf,y)}\right]$$

$$ICFVf,y = \frac{t_f}{2} \frac{\psi}{(\psi - \Delta\gamma_{,,}f)}$$

Once the *ICFVf,y* has been identified as a consequence of the predetermined loss of front *camber*, the consequent rear loss is determined, as follows.

$$\Delta\gamma_{,,}r = \chi\, \Delta\gamma_{,,}f$$

where χ is a number smaller than one that increases as the understeer gradient decreases. Carrying out all the calculations, we arrive at the determination of the position of the *ICFVr,y* as a function of *ICFVf,y* and χ:

$$ICFVr,y = \frac{t_r(ICFVf,y)}{2(ICFVf,y) - 2\chi(ICFVf,y) + \chi t_f}$$

Instead, the positions of the *ICFV,z* are evaluated through the heights of the roll centers according to the formula

$$ICFV,z = (ICFV,y)\tan\varphi$$

When deciding the value of the coefficient χ it must also be taken into account that the difference between the *camber* gains of the two axles must not be excessive due to the presence of *suspension jacking*. This phenomenon, which is more pronounced on the rear axle due to the greater height of the roll center, is not welcomed due to the consequent loss of *camber* in rebound travel which can lead to oversteer. When *suspension jacking* is significant, more gain in bump travel plays against it because, if the curve is symmetrical, it also means more loss in rebound. We shall now refer to the following figure, where only the effect of external forces ΔY is considered. Here, the lifting of the axle derives from the equilibrium between the moment of these external forces and the moment of the vertical forces due to the stiffness of the suspension. This displacement is superimposed onto the effects of the weight force and of the centrifugal force, which are absent in the following analysis. The application of the overlapping of the effects allows a good numerical prediction only in the case of small suspension travels, otherwise it provides only qualitative indications.

Automotive suspension

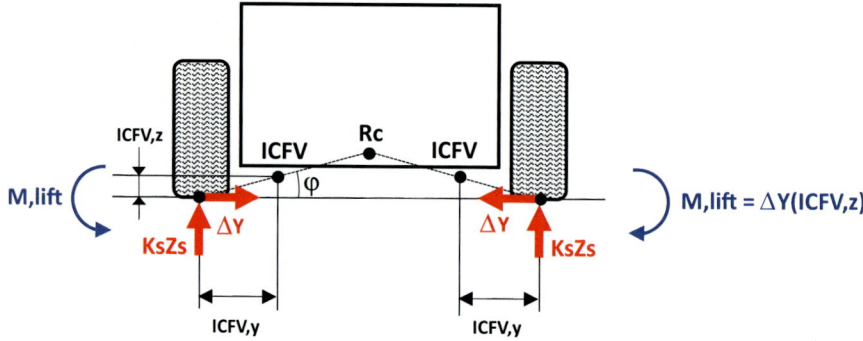

Figure 38 – Representation of the axle with the only forces that contribute to generating the lifting moment of the body of the vehicle.

The balance between the lifting moment and the moment of the ground reaction is:

$$KsZs(ICFV,y) = \Delta Y(ICFV,z)$$

In which Ks and Zs are respectively the stiffness in rebound travel and the extension of the suspension resulting from the lifting of the vehicle body. By carrying out the calculations, a qualitative assessment of the wheels rebound is obtained which depends on the height of the roll center, the lateral acceleration and the response of the tires to the variation in vertical load as described in chapter 2.

$$Zs = \frac{\Delta Y(ICFV,z)}{Ks(ICFV,y)}$$

$$Zs = \frac{\Delta Y}{Ks} \tan \varphi$$

If the CAD models of the suspensions are not available during the kinematics evaluation phase, the determination of the coordinates of the instantaneous centers becomes difficult because it requires the execution of measurements with a high degree of precision. In particular, with reference to the following figure, we can write:

$$D(ICFV - Ct) \Delta \gamma \, sen\varphi = dy$$

$$D(ICFV - Ct) = \frac{dy}{\Delta \gamma \, sen\varphi}$$

being:

$$ICFV,z = D(ICFV - Ct) \, sen\varphi$$

by replacing the previous expression, it results:

$$ICFV,z = \frac{dy}{\Delta \gamma}$$

The component y is obtained from the formulation of the *camber* gain in bump travel:

$$ICFV,y = \frac{dz}{\Delta \gamma}$$

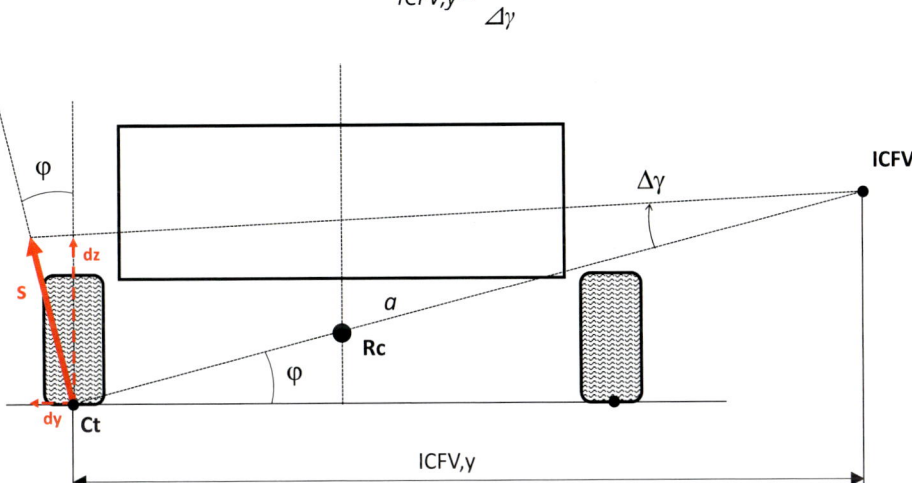

Figure 39 – Graphic construction of the displacement of the Ct point on the two axes for identifying the coordinates of the ICFV.

Therefore, from the measure of the wheel track variation and *camber* variation in bump travel it is possible to obtain the coordinates (with respect to the contact point at ground) of the *ICFV*.

6.7.4. Alteration of the kinematics in the roll transient due to the shock absorbers

The kinematics analyzed so far allows us to predict the behavior of the vehicle with good approximation only in stationary conditions. If, on the other hand, we are looking for an indication of what happens during transients, it is necessary to take into account the shock absorbers affecting the system only in the presence of a relative speed between the components. As shown in chapter 3, in some cases the presence of the shock absorbers can block some relative movements and vary the kinematics of a mechanism. A shock absorber with a very pronounced characteristic of extension compared to the compression one, locks the rebound (extension) of the inside wheel during the roll motion when cornering. Consequently, during the initial transient, the vehicle body rotates around the center of the contact patch of the inner wheel, thanks to the bump travel of the external wheel which gains *camber*. This happens due to the lateral elasticity of the tires, since the new kinematics obtained are isostatic. In the right part of the following figure there is a representation of what has just been described.

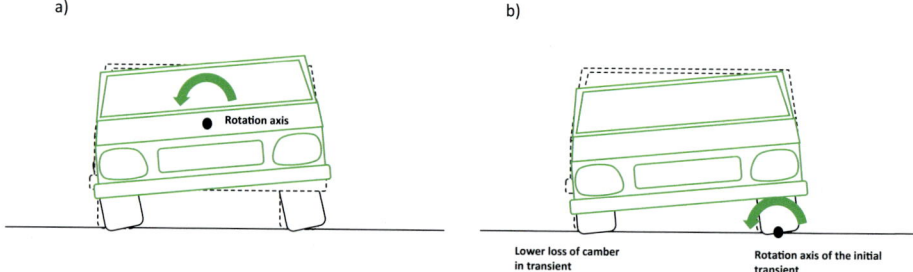

Figure 40 – a) Representation of the initial roll transient obtained with symmetrical (compression-rebound) shock absorbers. b) Representation of the initial roll transient obtained by increasing the extension characteristic of the shock absorbers with evidence of the lower loss of *camber* of the wheel outside the curve. Figure a) also represents the final stationary configuration of the roll which does not depend on the shock absorbers.

The demonstration of what has been stated is illustrated (only qualitatively because the elastic deformations are not taken into account) in the following figure. Here it is clear that the rotation around *Ct* produces, for the same roll angle, a smaller displacement *s* than the rotation around *Rc* of figure 36 of the previous paragraph, since the distance *Ct-ICSV* (*Ct* of the inner wheel) is smaller than the distance Rc-ICSV. The loss of *camber* of the wheel outside the curve that is obtained from the displacement *sp*, obtained by projecting the displacement *s* onto the perpendicular to the line joining the external point *Ct* and the ICFSV, is decidedly less than in the case of figure 36 in which the vehicle body rolls around at the roll center *Rc*.

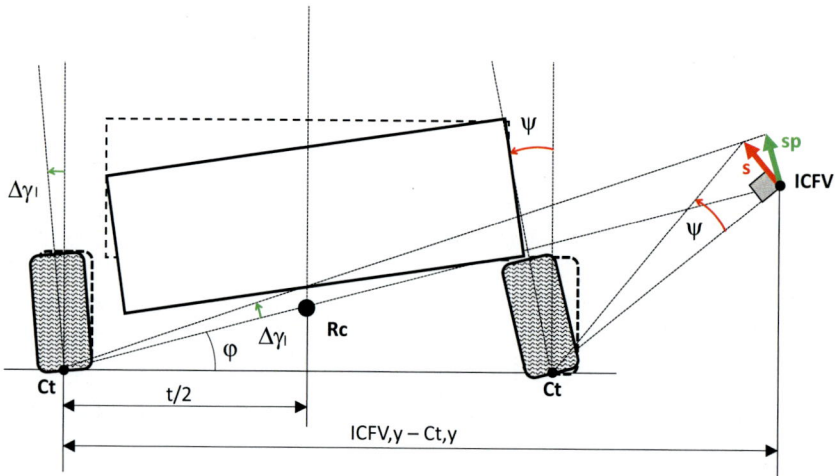

Figure 41 – Kinematic chain of the roll motion in the initial transient around the Ct inside the curve.

Given the above statements, we can say that a car with more braked rebound dampers works with more favorable average external *camber* angles. The exploitation of this phenomenon, if properly balanced between the axles, is useful in sudden changes of direction in which the more favorable *camber* angles make the car more reactive.

6.8. Longitudinal kinematics

Knowledge of the longitudinal kinematics of the suspensions is used to evaluate the contributions of the geometry of the two axles to the pitch of the vehicle under longitudinal load (traction and braking) and to identify the optimal location of the *ICSV*. We shall now see how to schematize this phenomena.

The kinematics projected on the longitudinal plane is comparable to a single-arm mechanism with a fulcrum in the *ICSV*, which in turn, in most architectures, has a variable position with the operation of the suspension. The movement of the wheel is thus a rotation around the *ICSV*, which represents an equivalent hinge with variable position in the motion of bump-rebound travel.

The longitudinal acceleration derived from braking produces a load transfer ΔFz and a longitudinal ground force Fx. Their resultant F, as can be seen in the following figure, forms an angle $\theta_{0.7}$ with the horizontal which, in adherence condition, is always constant and doesn't depend on the intensity of the deceleration. The moment of the force F with respect to the *ICSV*, representing the equivalent hinge, is Fd. This moment is the cause of the bump travel of the suspension under braking, and is canceled when the direction of the force F coincides with the straight line c obtained by joining the contact point at ground with the *ICSV*, that is when $\theta_{F,t} = \theta_{0.7}$. To adjust the pitching tendency, since the inclination of F is constant, we must act on the position of the *ICSV*, therefore on the kinematics.

% ANTI DIVE = (tan $\theta_{F,t}$/ tan $\theta_{0,7}$)100

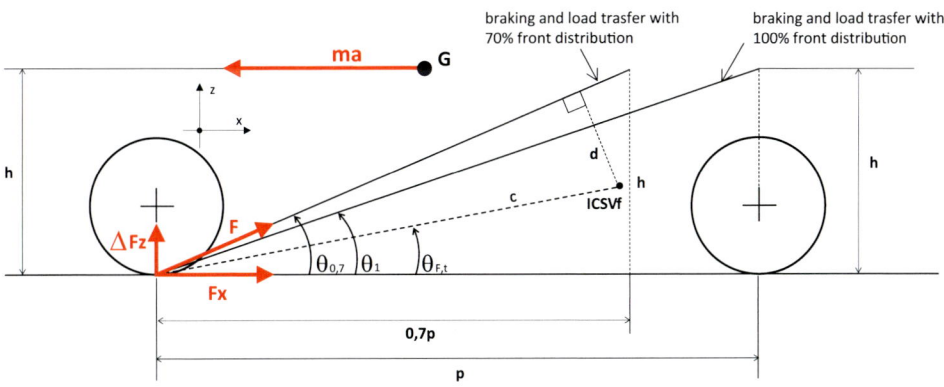

Figure 42 – Side view of the vehicle with evidence of the ICSV of the front suspension, the point of application and the direction of the braking force acting on the front axle.

The inclination of the force F turns out to be:

$$tan\theta_{0.7} = \frac{\Delta Fz}{Fx}$$

the vertical load transfer due to the longitudinal acceleration is:

$$\Delta Fz = \frac{mah}{p}$$

the brake force on the front axle is:

$$Fx = maf_a$$

by replacing

$$\tan\theta_{0.7} = \frac{mah}{pmaf_a} = \frac{h}{pf_a}$$

in which f_a is the number less than 1 which represents the brake partition on the front axle which in our case we imagine is 0.7. The subscript used for θ indicates just that.

The *anti dive* percentage is the ratio between the inclination of the equivalent hinge $\tan\theta_{F,t}$ and the inclination of the force F. This ratio is always less than 1 because in general, in order to have a non-zero bump travel value during braking, the force F is always more inclined than the equivalent hinge. The formulation of the *anti dive* percentage is as follows:

$$\%ANTI\ DIVE = \frac{\tan\theta_{F,t}}{\tan\theta_{0.7}}\ 100$$

Substituting the expression of $\tan\theta_{0.7}$ we get:

$$\%ANTI\ DIVE = \frac{\tan\theta_{F,t}}{h/(pf_a)}\ 100$$

The inclination of the force F is the acute angle at the base of the right triangle obtained by raising from the ground line, at a distance from the center of the front wheel equal to the pf_a, a vertical segment of length h and joining its end with the front point at ground, as shown in the previous figure.

One hundred percent *anti dive* means that the direction of the force F passes through *ICSV* while zero percent means that *ICSV* is on the ground line.

In general, whatever the mission of the vehicle, it is always preferable to have little bump under braking in order not to have excessive vehicle height variations that can be annoying both for aerodynamic reasons and because the bump travel of the suspension leads to variations in the *characteristic angles* (*camber* and *toe*) that could affect braking efficiency, as they lower the adherence capital of the tire in longitudinal direction. Also, the bump in braking represents a delay in the vertical load transfer between the two axles, that is, the achievement of the optimal braking condition. The upper limit lies in the fact that as the *anti dive* percentage increases, the effect of the braking force on the vehicle body is more abrupt. This is because it does not pass through the spring which softens the reaction. Furthermore, the positioning of the *ICSV* (quite high), which requires one hundred percent *anti dive* geometry, produces annoying wheelbase variations in bump

travel that are generally undesirable both because they lower *comfort* and because they induce the axles to assume self-steering configurations. In some grand touring cars the *θF,t* inclinations vary from 2 to 4 deg and the *anti dive* percentages from 10 and 15%.
As for the rear axle, the situation is completely similar and is represented in the following figure.

% ANTI LIFT, rear = (tan $\theta_{R,t}$/ tan $\theta_{0,3}$)100

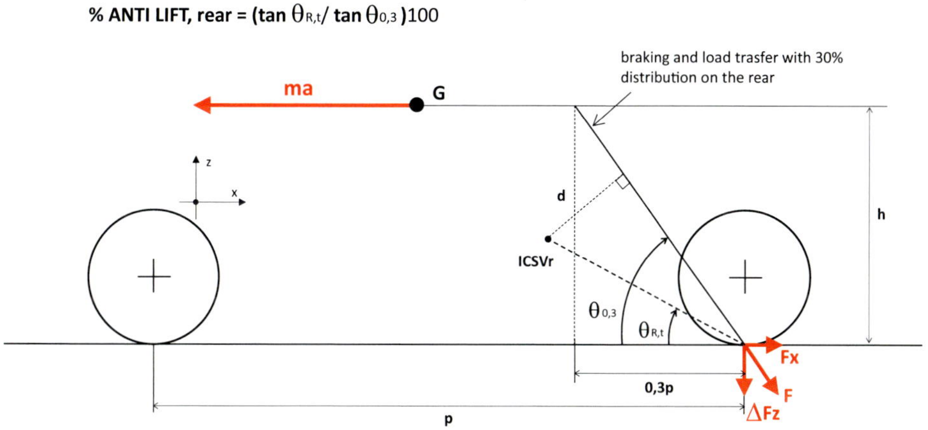

Figure 43 – Longitudinal section of the vehicle showing the Instant Center Side View (ICSVr) of the rear suspension and the direction of the braking force acting on the rear axle.

In which it results:

$$\%ANTI\ LIFT, rear = \frac{tan\theta_{R,t}}{h/[p(1-f_a)]} \cdot 100$$

and then:

$$\%ANTI\ LIFT, rear = \frac{tan\theta_{R,t}}{tan\theta_{0.3}} \cdot 100$$

The inclination of the force *F* is the acute angle at the base of the right triangle obtained by raising from the ground line, at a distance from the rear wheel center equal to *p(1-fa)*, a vertical segment of length *h* and joining its end with the rear point at ground.
The same goes for the tendency to pitch in traction, with the difference that the characteristic is called *anti lift*, in the case of front-wheel drive, and *anti squat* in the case of rear-wheel drive. The traction force in the case of drive-shafts transmission must be applied to the center of the wheel while in the case of chain transmission it must be applied to the ground. In the following figure the graphic representation of the front drive-shafts transmission case is shown.

Automotive suspension

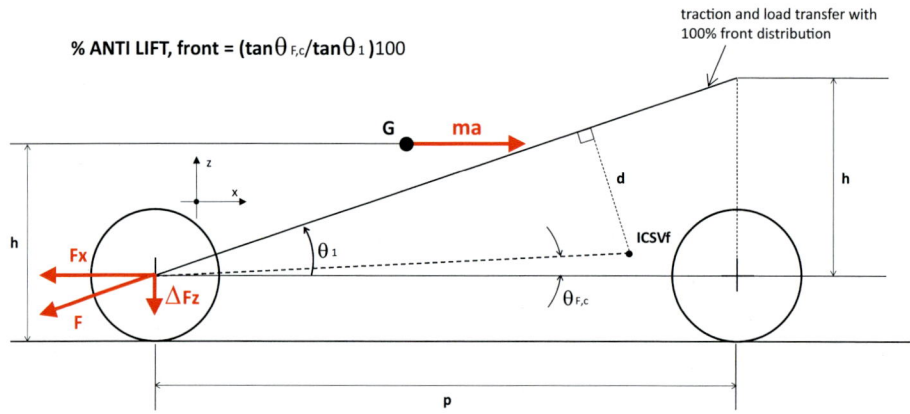

Figure 44 – Side view of the vehicle with evidence of the Instant Center Side View (ICSVf) of the front suspension and the direction of the resulting traction force acting on the front axle in the case of front-wheel drive.

In this case the value of inclination of the resultant force with respect the horizontal is:

$$tan\theta_1 = \frac{mah}{pmat_a} = \frac{h}{pt_a}$$

where *ta* is a number less than 1 which represents the distribution of the traction and its value is 1 when the car is front-wheel drive. In four-wheel drive cars the inclination of *F* increases.

The *anti lift* percentage:

$$\%ANTI\ LIFT, front = \frac{tan\theta_{F,c}}{tan\theta_1} 100$$

by replacing the expression of *tan θ₁*

$$\%ANTI\ LIFT, front = \frac{tan\theta_{F,c}}{h/(pt_a)} 100$$

It is very important that the *ICFSVf* is higher than the wheel center in order to lower the lifting moment *Fd* due to traction. Unfortunately, this goes against *comfort* because, in this configuration, the car increases the wheelbase (at wheel center) in bump travel and the wheel tends to move against an obstacle.

With reference to the following figure, in the case of rear-wheel drive, the tendency to squat becomes:

$$\%ANTI\ SQUAT = \frac{tan\theta_{R,c}}{h/(pt_p)} 100$$

where *tp* is a number less than 1 that represents the distribution of traction and that when the car has rear traction it is 1. From the formula it is clear that when the traction is also distributed towards the front (*tp* is <1) the percentage of *anti squat* decreases and the vehicle tends to squat more.

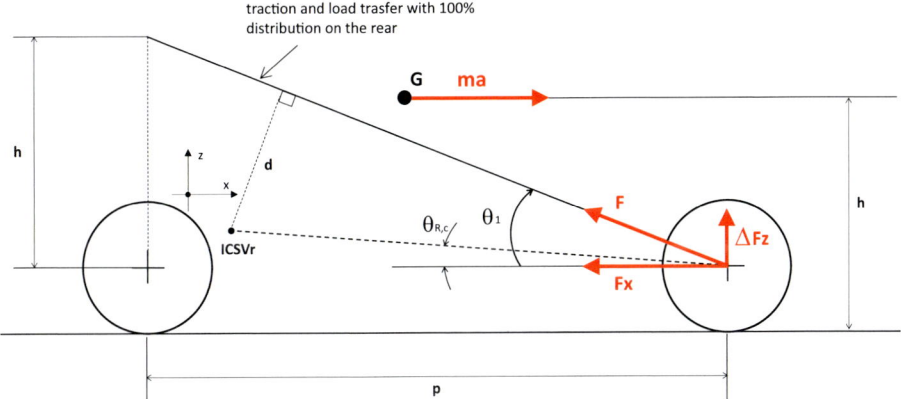

% ANTI SQUAT = $(\tan\theta_{R,c}/\tan\theta_1)100$

Figure 45 – Side view of the vehicle showing the Instant Center Side View (ICSVr) of the rear suspension and the direction of the traction force in the case of rear-wheel drive.

In rear-wheel drive cars, squatting represents a delay in the increase in load on the rear axle, that is, in reaching a better configuration for traction.
Summarizing the most important formulas we have:

Brake

$$\%ANTI\ DIVE = \frac{\tan\theta_{F,t}}{h/(pf_a)}\ 100$$

$$\%ANTI\ LIFT, rear = \frac{\tan\theta_{R,t}}{h/[p(1-f_a)]}\ 100 \quad (f_a=0.7)$$

Traction

$$\%ANTI\ LIFT, front = \frac{\tan\theta_{F,c}}{h/(pt_a)}\ 100 \quad (t_a=1\ front\text{-}wheel\ drive;\ t_a <1\ four\text{-}wheel\ drive)$$

$$\%ANTI\ SQUAT = \frac{\tan\theta_{R,c}}{h/(pt_p)}\ 100 \quad (t_p=1\ rear\text{-}wheel\ drive;\ t_p <1\ four\text{-}wheel\ drive)$$

Automotive suspension

It is very useful, from a practical point of view, to evaluate $\theta F,c$ and $\theta F,t$ by measuring the variation of the wheelbase at ground and at wheel center of the front suspension as a function of the bump travel because, with reference to figure, it turns out to be:

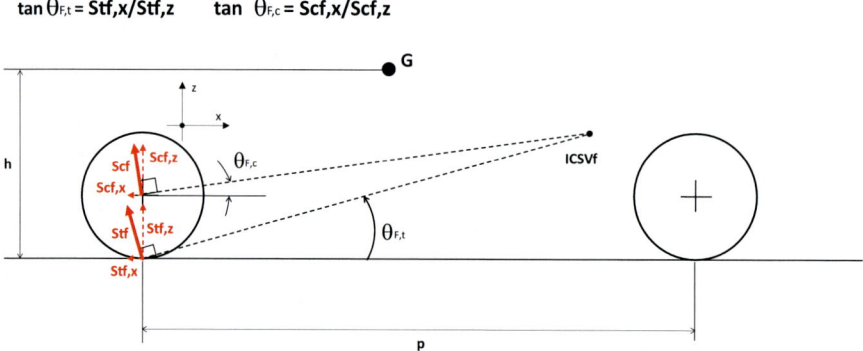

Figure 46 – Side view of the vehicle with the Instant Center Side View (ICSVf) of the front suspension and the wheelbase variations at the ground and at wheel center.

In which $tan\theta F,t$ is positive when, during the bump travel of the wheel, the wheelbase variation is positive (the ground point advances) while $tan\theta F,c$ is positive when the wheelbase variation at wheel center is positive (the wheel center advances) in the same maneuver.

If Str e Scr are respectively the displacements on the ground and at the rear wheel center, that have not been shown in figure in order not to overcomplicate it, the formulas became:

Brake

$$\%ANTI\ DIVE = \frac{Stf,x/Stf,z}{h/(pf_a)}\ 100$$

$$\%ANTI\ LIFT, rear = \frac{Str,x/Str,z}{h/[p(1-f_a)]}\ 100 \quad (f_a=0.7)$$

Traction

$$\%ANTI\ LIFT, ant = \frac{Scf,x/Scf,z}{h/(pt_a)}\ 100\ (t_a=1\ front\text{-}wheel\ drive;\ t_a <1\ four\text{-}wheel\ drive)$$

$$\%ANTI\ SQUAT = \frac{Scr,x/Scr,z}{h/(pt_p)}\ 100\ (t_p=1\ rear\text{-}wheel\ drive;\ t_p <1\ four\text{-}wheel\ drive)$$

6.8.1. Calculation of the ICSV position

In analogy with what has been explained for the *camber* variation in bump travel, the longitudinal position of the *ICSV* influences the relationship between the rotation of the upright around an axis transverse to the vehicle (parallel to y and passing through the *ICSV*) and the vertical displacement of the wheel center. This rotation, called *spin*, must be controlled because it involves a change in the configuration of the suspension with consequent variation of other quantities. The *spin* variation does not always coincide with the *caster* variation because in some suspensions, such as the McPherson, the longitudinal and vertical displacement of the control arm ball joint, consequent to the upright rotation in bump travel, produces a further *caster* variation, since the center of the top mount does not move. In the following figure, the representation of the *spin* angle in a McPherson type front suspension in which the advancement of point A (ball joint of the control arm) depends on the fact that the *ICSV* is higher than the latter. By positioning the *ICSV* below point A (moving it along the vertical), with the same Δ*spin*, the retreat of ball joint is obtained, and consequently a lower caster variation is obtained. For example, a kinematics with an equal and opposite angle φ (negative), with respect to that of the figure, and with the same distance between the *ICSV* and point A, with the same Δ*spin*, has the same Sz but opposite Sx (retreat) and consequently a lower slope of the caster variation with respect to vertical displacement.

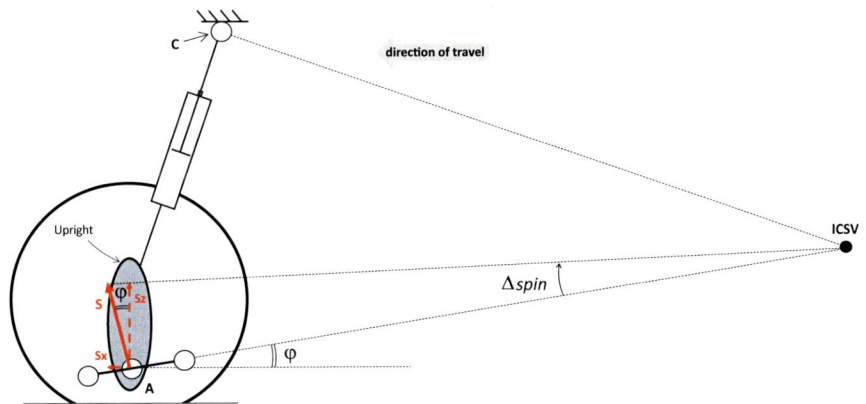

Figure 47 – Kinematics of a McPherson type front left suspension projected on the longitudinal plane passing through the wheel center with the projections, in x and y, of the displacement S of the center of the ball joint of the control arm. With the same Δ*spin* and S, the Sx displacement depends on the height of the ICSV.

To identify the *ICSV* position, we can start by assuming that the maximum *spin* variation is equal to the maximum caster variation, the values of which are described in chapter 7. Subsequently, the first optimization of the longitudinal and vertical positions of the *ICSV* is carried out by verifying, through the virtual analysis, that the actual *caster* variation is aligned with the mission of the vehicle. To modulate this last variation, we intervene, for example, on the trend of the longitudinal displacement of the ball joint of the control arm or on those quantities which, from the analysis of the kinematics, cause further variations in the position of the steering axis. We shall now evaluate the *spin* variation with respect to the bump travel of the front suspension which, similarly to the *camber* variation in bump travel, follows the formula:

$$\frac{\Delta spin}{dz} = \frac{1}{ICSVf,x}$$

In which the *ICSVf,x* is the distance, along x, between the *ICSVf* and the center of front contact patch. Knowing therefore, the maximum variation of *spin* in radians, the position of the *ICSVf,x* is calculated as follows:

$$ICSVf,x = \frac{1}{\left(\frac{\Delta spin}{dz}\right)}$$

By overlapping the trends to dive under braking and to lift under acceleration we can build a schematization to graphically identify the optimal z of the *ICSV*, as shown in the following figure.

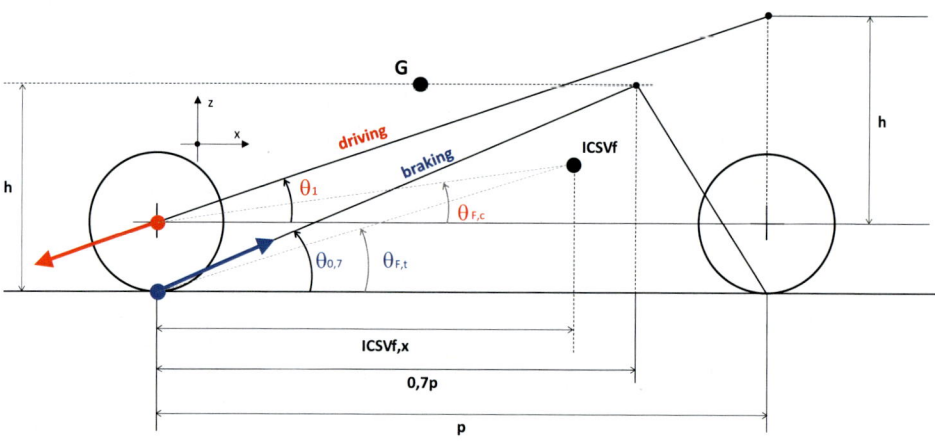

Figure 48 – Longitudinal section of the vehicle with the superimposition of the quantities relating to *anti dive* under braking and *anti lift* under acceleration for the identification of the possible positions of the front ICSV, in a front-wheel drive vehicle.

As a first attempt, the *ICSVf* can be positioned below the *ICFVf* (for *toe-out* in bump travel) and below the two straight lines of action of the braking and traction in order to have a good percentage of *anti dive* (ratio between tangents of the angles in blue) and a good percentage of *anti lift* (ratio between tangents of the angles in red). In this choice, the consequent wheelbase variation at wheel center and at ground must be taken into account. As already mentioned, to have more *comfort* when passing an obstacle, the wheelbase variation at wheel center must be as small as possible. This so that the wheel, in its bump travel motion, does not advance too far towards the obstacle. However, this also produces a lowering of the aforementioned percentages. In rear-wheel drive cars, since there is no *anti lift* requirement on the front axle, the strategy of positioning the *ICSVf* below the wheel center is followed with the aim of having the wheel retracting during the obstacle passage. In this case, however, the tendency to pitch during braking increases.

It shall be recalled that, with reference to figure 46, the wheelbase variation at wheel center follows the formula:

$$\frac{Scf,x}{Scf,z} = tan\theta_{F,c}$$

while the wheelbase variation at ground follows the formula:

$$\frac{Stf,x}{Stf,z} = tan\theta_{F,t}$$

The first members of the above formulas are always known as objectives to be pursued. In particular, if the goal is *comfort*, the angle $\theta_{F,c}$ is calculated from the first formula, starting from the desired wheelbase variation (at wheel center) that allows it. The consequent $\theta_{F,t}$ is then calculated. Otherwise, if we aim for the maximum *anti dive* percentage, the *ICSVf* must be positioned as close as possible to the braking action line (always very inclined) and the angle $\theta_{F,t}$ must be calculated tolerating a significant increase in the wheelbase variation, both on the ground and the center of wheel. The limit of this strategy lies in the fact that, by allowing important ground wheelbase variation equal to 10mm on 100mm di bump travel, it is possible to obtain *anti dive*% that do not exceed 30%. This is because the inclination $\theta_{0.7}$ of touring cars is never low but always remains around 20 degrees (p = 2600 mm, h = 650 mm, *fa* = 0.7), while to obtain the aforementioned pitch variation a $\theta_{F,t}$ of approximately 6 degrees is sufficient. Therefore, if we want to increase the percentage of *anti dive* we must accept more wheelbase variation on the ground and the consequent negative effects on *comfort* due to the increase of the wheelbase variation at wheel center. In addition, an excessive change in wheelbase produces longitudinal sliding which compromises the effectiveness of braking.

For the rear axle the situation is similar. In the following figure, the overlap between the quantities that produce squatting in traction and those that produce lifting when braking is shown.

Automotive suspension

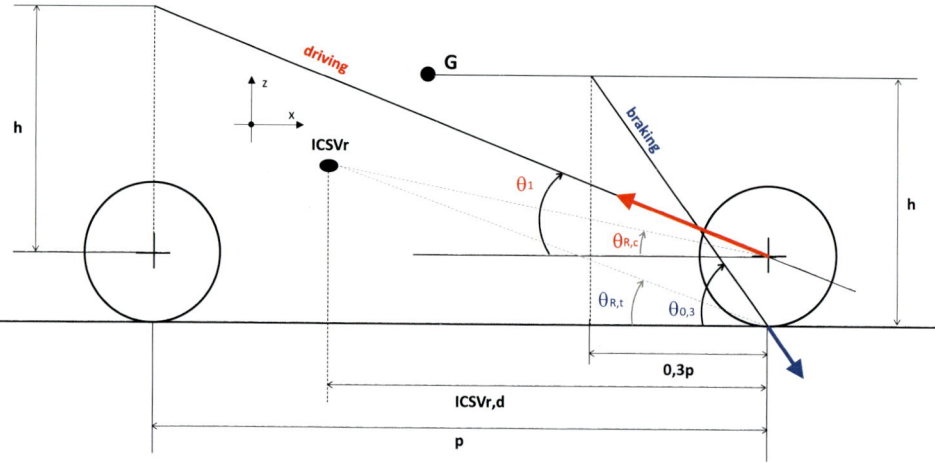

Figure 49 – Longitudinal section of the vehicle with the superimposition of the quantities relating to the *anti lift* under braking and *anti squat* under acceleration for the identification of the possible positions of the rear ICSV, in a rear-wheel drive vehicle.

Following the same methodology shown for the front axle, it is possible to calculate the distance of the *ICSVr* from the rear axle with the aim of the variability of the *spin* angle equal to the maximum variation of *caster* which, in this case, must be more contained (within 1 degree on 100mm of bump travel). By inserting this last variation in radians in the following formula, the desired distance is obtained.

$$ICSVr,d = \frac{1}{\left(\frac{\Delta spin}{dz}\right)}$$

As for the height, the first step is to position the *ICSVr* below the *ICFVr* to obtain *toe-in* in bump travel. To limit the *anti squat* and *anti lift*, the *ICSVr* must be brought closer to the braking and traction lines but, unlike the front axle, this improves the *comfort* because the wheelbase variations tend to increase. When we decide to maximize traction, it is necessary to bring the ICSVr as close as possible to the line of action of the traction. This way, the bump travel in acceleration is limited and the variation of the characteristic angles (*camber* and *toe*), which generally favor the loss of grip, is reduced. Usually, the goals for defining the height of the *ICSVr*, which consist of *toe-in* in bump travel, in the wheelbase variation that regulates the *comfort* and in the maximum percentage of *anti squat*, are in contrast because the achievement of the first precludes the reaching the other two and vice versa.

In sports cars where the center of gravity is very low, since the braking and traction lines of action are less inclined, it is easier to average the objectives because the *ICSVr* can be approached to these lines with less risk of disturbing the *toe* variation and without increasing the wheelbase variation too much. This last statement applies to both the front and rear axles.

7. THE STEERING SYSTEM

The steering system is an integral part of the front suspension and is the mechanism that allows the car to turn thanks to the steering of the front wheels around their respective axles. The position of the steering axis of a suspension defines the first six geometric quantities mentioned in the previous chapter. The steering mechanism allows control of the relative steer of the two wheels which, with imposed steering wheel angle, determines the trajectory and response of the vehicle when changing direction. The following description will start from the inclination of the steering axis to conclude with the kinematics with all its characteristics.

7.1. The *Caster*

The *caster* or upright incidence is the angle between the projection of the steering axis on the longitudinal plane of the vehicle and the vertical. To define the *caster*, it is necessary to identify the steering axis, that is, the axis around which the wheel moves during steering. It is the joining between the centers of the kinematic constraints of the upright with respect to the arms or to the body of the vehicle, depending on the type of suspension. To better clarify this definition, we must refer to the following figure where a typical front suspension architecture called Double wishbone is schematized. In this case, the upright is linked to the upper arm and the lower arm by means of two ball joints and rotates around the joining between their centers which thus determines the steer axis. The rotation is induced by the thrust of the steering rod which, in turn, is constrained to the upright by means of another ball joint.

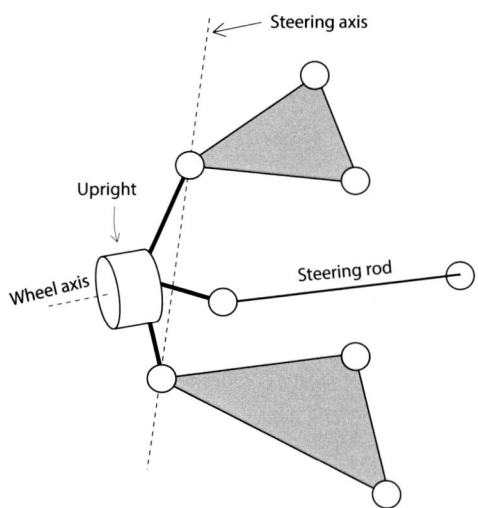

Figure 1 – Schematic of a front Double wishbone – left side.

Automotive suspension

The following figure shows, a MPherson scheme in which the upright is connected to the arm with a ball joint and is integral with the shock absorber, which in turn is linked to the vehicle body with a ball joint. The steering axis is the connecting line between the centers of the latter joints.

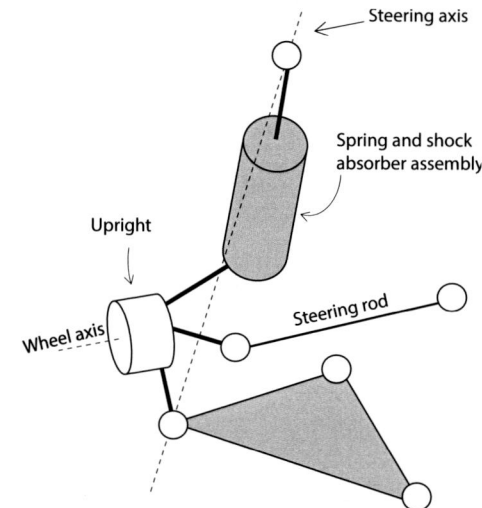

Figure 2 – Schematization of the McPherson front suspension – left side.

By projecting the steering axis on the longitudinal plane, we arrive at the following schematic where both the *caster* and *caster trail* (the longitudinal arm at ground) are represented.

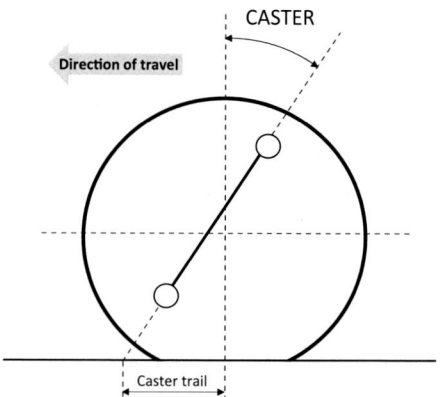

Figure 3 – *Caster* angle in a side view.

The *caster* is positive when the trace of the steer axis on the road plan is advanced with respect to the center of the contact patch. This condition, which is the only most used one, favors greater stability because the accidental loss of directionality of the wheel is recovered thanks to the self-alignment induced by the force of resistance to advancement, as can be seen in the following figure. This *first stabilizing action* (we will call this influence this way) is indirect, as it does not depend on the *caster* but on the positive value of the *caster trail* (longitudinal arm at ground) due to the *caster*. The *caster trail*, which will be discussed in detail in the next paragraph, is a consequence of the *caster*. However, it can also exist with a null *caster* and forward steering axis with respect to the center of the wheel, as happens for the shopping cart. The amount of this first stabilizing action is linked to the resistance to the advancement which, starting from an initial low coefficient (0.01-0.015), undergoes a much more pronounced increase as a function of speed on road tires than on sports tires (with greater performance requests). It follows that the first stabilizing action, generally effective only in the presence of small external disturbances, can only increase its effect with road tires which, at high speed, can reach resistance coefficients equal to 0.03-0.04.

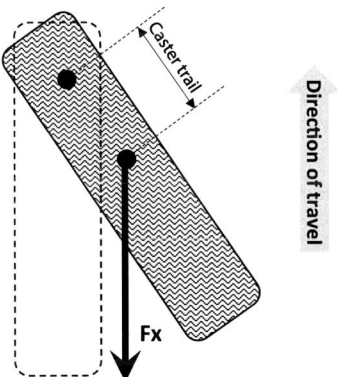

Figure 4 – Plan view of the wheel with trace of the steering axis and *caster trail* in the case of a positive *caster* with evidence of self-alignment.

A very important feature of the positive *caster* is the *camber* gain in steering. This phenomenon can be quickly explained by looking at the following figure in which, by rotating the wheel by 90 degrees, we obtain a *camber* angle equal to the *caster*.

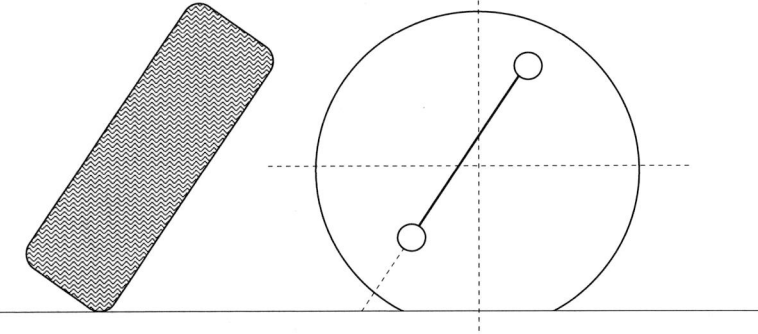

Figure 5 – On the right the lateral view of the steering axis of the left side, on the left the lateral view of a wheel that has undergone a rotation of 90 degrees around the steering axis in a clockwise direction.

From this it can be seen that the more the *caster* increases, the more the *camber* gain in steering increases and therefore the slip stiffness of the front axle when cornering. To allow greater *camber* gain in cornering, the *caster* variation in bump travel (compression of the spring) is also controlled. In road cars, it takes values around 1.5 degrees per 100 mm, while in grand touring and racing cars it assumes values around 2-3 degrees for 100 mm of bump travel. It is necessary to carefully weigh the value of the *caster* and its variations as a function of the *camber* gains on the two axles to always have a concord $\Delta\phi$. A too high *caster* value, since increases the slip stiffness of the axle, could make the car oversteer and have a discord $\Delta\phi$ which is what happens in drifting cars.

The law that links the steering wheel angle to the *camber* angle is used to measure the *caster* on the benches for measuring the vehicle wheel angles. To carry out this measurement, steerings are carried out both to the left and to the right which produce a set of *camber* angles which, together with the corresponding steer angles on the wheels, are inserted into a mathematical formula for calculating the *caster*. In this measure, it is very important to consider that, due to the hysteresis of the vehicle, the resulting *camber* and therefore the *caster* depend on the speed at which the steering is made. It is therefore suggested to carry out the steering cycle at the minimum possible speed so as not to contaminate the measurement.

Another possible contamination of the measurement depends on the fact that the inclination, both longitudinal and transversal of the steering axis, causes the vehicle to lift when cornering because the contact points tend to penetrate the road plan as they are forced to carry out a trajectory perpendicular to the steering axis. This phenomenon affects not only the torque at the steering wheel, but also the measure of the *caster*, because the resulting *camber* angles are affected by the change in slope of the car due to the asymmetrical lifting. In the following figure the graphic explanation of the phenomenon is shown.

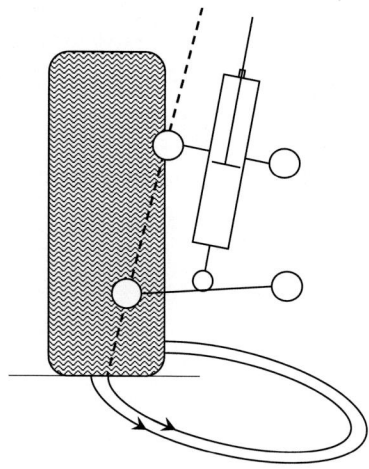

Figure 6 – Trajectory of the contact points on the ground that tend to penetrate with the consequent asymmetrical lifting of the vehicle and the variation of the *camber* with respect to the vertical.

The phenomenon of the car lifting produces the *second stabilizing action* (as we will call this influence), attributed directly to the *caster*, which is exploited when the amount of external disturbances becomes significant. The *second stabilizing action* arises thanks to the elastic return of the external wheel in the previous position due to the increase in the vertical load. This manifests itself with the increase of steering torque which usually is absorbed by the steering power or by the user in the absence of it. The increase in vertical load, which is partly dampened after the car is lifted, in the first transient contributes to giving greater slip stiffness to the front axle.

The values of the *caster* angle range from 2.0-2.5, the minimum values adopted for commercial and utility vehicles, to 6-7.5, adopted for racing and sports cars. Usually, the main effect that is sought when increasing the *caster* is a greater *camber* gain in steering to increase the slip stiffness of the front axle and make the vehicle more direct and faster when changing direction. Attention should be paid to the fact that when the *caster* is increased, with the same torque to the steering wheel set by the user, the car has greater resistance to cornering, making it less agile because it is more stable, while with the same steering wheel angle it is faster. Therefore, if the *caster* angle is increased, it is also necessary to increase the power steering, otherwise the car may seem more rigid in the change of direction.

On rough terrains, where disturbances are more frequent and the *first stabilizing action* loses influence, by increasing the *caster* angle, the directional stability of the wheels can be improved and annoying and dangerous vibrations at the steering wheel can be avoided; for this reason, *caster* angles that exceed 5 degrees are also used in off-road vehicles in order to exploit the *second stabilizing action*. Also, to improve stability, rear-wheel drive cars have higher *caster* angles than those of the same segment with front-wheel drive. The reason lies in the fact that front-wheel drive is equivalent to a pair of symmetrical longitudinal forces, applied near the wheel centers. This has a stabilizing effect on the steering because, having fairly high values, it tends to make the effect of external perturbations negligible. Even on cars with mass distribution shifted towards the rear, it is necessary to increase the *caster* angle to increase the effect of the *second stabilizing action* which is proportional to the weight. In the latter type of cars, the increase in the *caster* angle also serves to improve the yawing capacity (lateral force with the same set steering angle) of the front axle, which may not be adequate due to the contained vertical load.

Automotive suspension

7.2. The *Caster trail*

The *caster trail* is the distance, in the longitudinal direction, between the trace of the steering axis on the ground and the projection of the wheel center on the ground. In the following figure, are shown the representations of the *caster trail* in the case of a *caster* other than zero and in the case of a *caster* equal to zero (shopping cart); the latter situation was proposed only for educational purposes but never occurs.

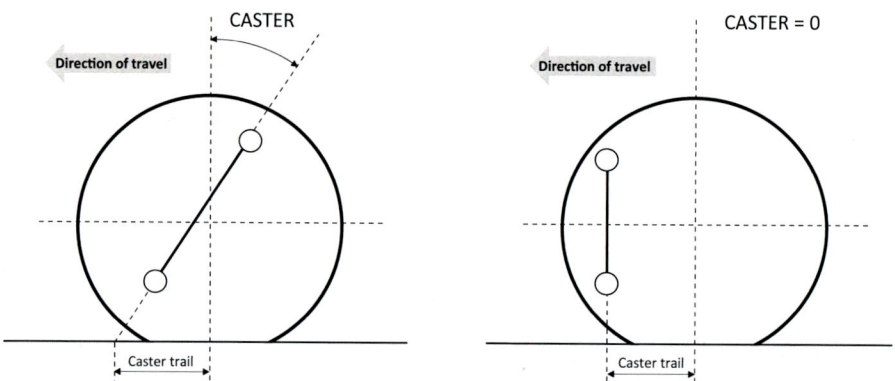

Figure 7 – *Caster trail* in a side view of the vehicle in the case of a *caster* other than zero and in the case of a *caster* equal to zero (shopping cart).

The *caster trail* greatly influences the divergence (wheel opening) under lateral load and is the most influential quantity on the *steering feeling* when cornering. This is because the reaction moment at the steering wheel, which guarantees the user the necessary *feedback* on the adherence of the car, depends on caster trail and on the lateral forces at ground. This last statement can be explained by the following figure, which shows the plan view of the front axle intent on turning to the right.

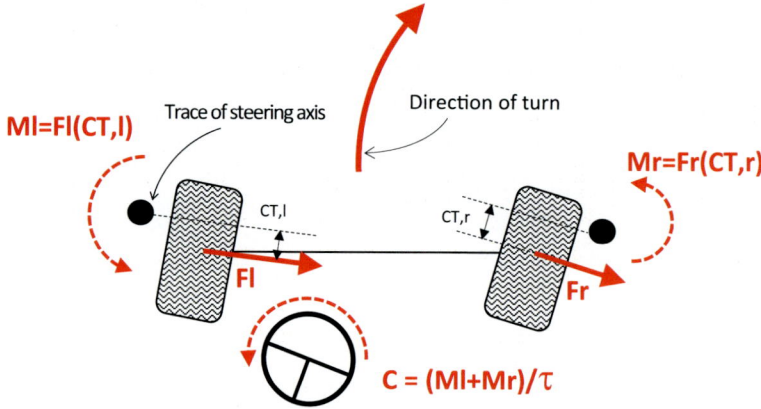

Figure 8 – Plan view of the front axle showing the reaction torques Ml and Mr on the wheels and C on the steering wheel, in a curve maneuver.

The lateral reaction forces *Fl* and *Fr*, due to the presence of the two *caster trails CT,l* and *CT,r*, produce the two reaction torques *Ml* and *Mr* which go up to the steering wheel with the reaction torque *C* equal to the quotient between (*Ms* + *Md*) and the steering ratio τ. The torque at the steering wheel *C* represents the *steering feeling* or the *feedback* that reaches the user on the adherence conditions from which the two reaction forces *Fs* and *Fd* on the tires arise. If, with the same *CT,l* and *CT,r*, the forces *Fl* and *Fr* suddenly drop, the consequent decrease in the torque *Ml* and *Mr* and the torque on the steering wheel *C* represents the *feedback* to the user of the lesser adherence. The torque *C* takes the name of *dynamic torque* and represents only one of the three contributions of the total torque at the steering wheel. The second contribution is represented by the gravitational torque, due to the lifting of the car in steering. The third contribution by the torque due to friction, also known as the *static steering torque*, which depends on the characteristics of the tire and lowers considerably when the car is in motion. The ideal condition is when the *dynamic torque* has a fairly linear increase with lateral acceleration with a slight lowering just before reaching the *sliding limit* of the front tire. The tuning of the *dynamic torque*, which is the most adjustable of the three, is of fundamental importance because it represents one of the main indicators of the vehicle behavior towards the user. The achievement of the linear trend of *dynamic torque* presupposes that the *caster trails* (left and right) remain as much as possible constant during the operation of the suspension in curve (bump-rebound travel and steering). In this way the *feedback* towards the user (torque *Ml* and *Mr*) depends only on the trend of the lateral characteristic of the front tire and the reaction torque at the steering wheel assumes a constant slope for small lateral acceleration and decreasing near the *sliding limit* of the tire. The behavior described allows the user to understand when the axle is close to the crisis. If, on the other hand, the two *caster trails* (left and right) undergo significant variations during the operation of the suspension, the *dynamic torque* trend assumes an undesired behavior. It could even decrease with lateral acceleration, giving the user back the lightening of the steering wheel, which generally is a *feedback* interpreted as loss of adherence. The usual values of *caster trails* range from 5 to 25 mm. The low values are used for small cars where there is a tendency to lower the torque at the steering wheel in order not to adopt important power steering. The highest values are adopted in cars that aim to improve the *dynamic torque* performance and the resulting *steering feeling*. This is because, starting from a high static value, the variations are less influential and it is more difficult to generate the unwanted change of sign that produces the sudden lowering of the steering wheel torque. On the other hand, the negative effects of high values of *caster trail* are the following:

- Steering wheel vibrations on rough roads: as shown in the following figure, as the *caster trail* increases the moments of the lateral forces, deriving from the asperities, increase. The latter go up to the steering mechanism causing steering wheel vibrations. This effect exists only when the angle of incidence is low, and the *caster trail* value depends on the fact that the steering axis is advanced with respect to the wheel center. Otherwise, as the *caster* increases, the stabilizing effect increases (*second stabilizing action*), which contrasts with the lateral perturbations, and the vibration is resized until it disappears.

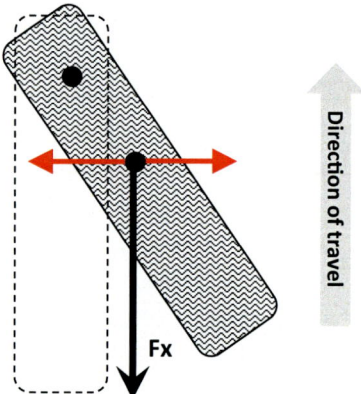

Figure 9 – Lateral forces acting on the wheel due to road roughness.

- Lateral wind gust: from the following figure it can be seen that a lateral thrust, resulting from a wind gust, produces the two opposite lateral forces on the front tires that cause the wheels to turn in the direction of the wind. Thanks to the increase of the *second stabilizing action*, this phenomenon also decreases as the *caster* increases.

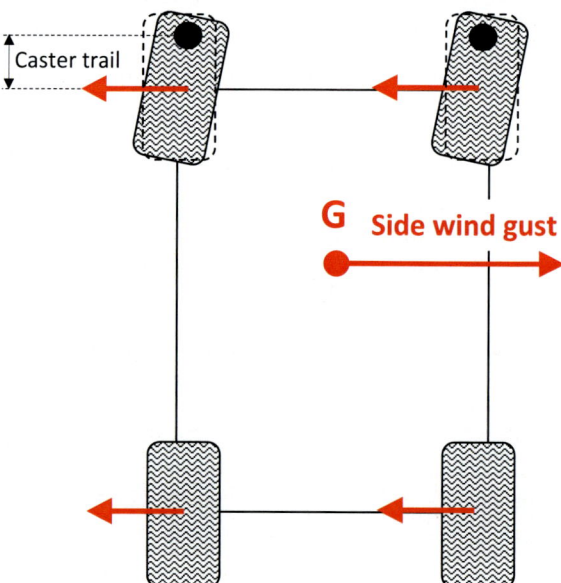

Figure 10 – Schematization of the front forces emerging in a side wind gust with evidence of the steering of the front wheels.

7.3. The longitudinal arm at wheel center

The longitudinal arm at wheel center or *Steering Axis Offset* (*SAO*) is the distance, in the longitudinal direction, between the steering axis and the wheel center as described in the following figure.

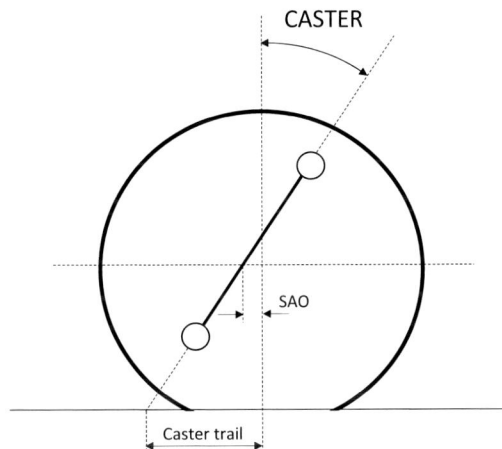

Figure 11 – Longitudinal arm at wheel center.

Generally, the *SAO* values for front-wheel drive cars are close to zero to limit the sliding of the drive shaft and prevent it from slipping out. As the *SAO* increases, the amount of *caster trail* which is not dependent on the *caster* and that, on rough terrain, produces the vibratory and instability effects seen in the previous paragraph increases. The goal is to keep this quantity close to 0 as much as possible but, being dependent on the *caster* and the *caster trail*, it is considered more important to keep the latter under control and let the *SAO* develop accordingly. In some situations, negative *SAO* values might be adopted to lower the amount of transverse actions that cause steering wheel vibrations (Figure 9).

7.4. The King Pin Inclination

The *KPI* is the angle between the projection of the steering axis on the transverse plane of the vehicle and the vertical as described in the following figure. Generally, the *KPI* tends to be lowered as much as possible because, unlike the *caster*, it produces a loss of *camber* when steering but is difficult due to space constraints. Referring to the following figure, similarly to what we saw for the *caster*, to obtain a *camber* value equal to the *KPI* we need to rotate it by 180 degrees. Therefore, the overall gain of *camber* in steering is positive because the loss due to the *KPI* is less significant than the gain due to the *caster* and to increase this effect we tend to lower the *KPI* as much as possible. The used values range from approximately 10 to 13 degrees.

Automotive suspension

7.5. The King Pin Offset

With reference to the following figure the *King Pin Offset* (*KPIoffset*) is the distance, along the transverse direction y, between the steering axis and the wheel center at the height of the wheel center itself.

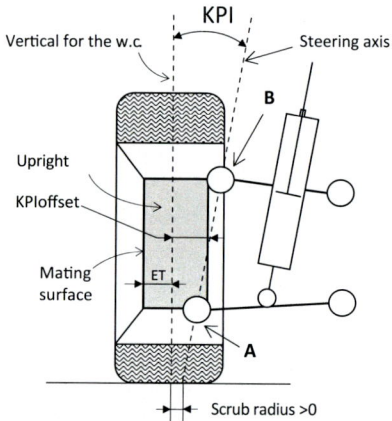

Figure 12 – Double wishbone with evidence of *KPIoffset* and the rim offset.

The value of *KPIoffset* must be contained (< 60 mm) to avoid the vibrations at the steering wheel in case of asymmetrical obstacle passing (as happens, on racetrack, when taking the curb on a curve). The load condition that simulates the asymmetrical obstacle passing, described in the following figure, is a variable longitudinal force *F* applied to the wheel center, in this case on the left side, which tends to produce vibrations on the steering wheel due to the moment F(KPIoffset). The more this distance decreases, the more the moment and the consequent vibrations on the steering wheel decrease.

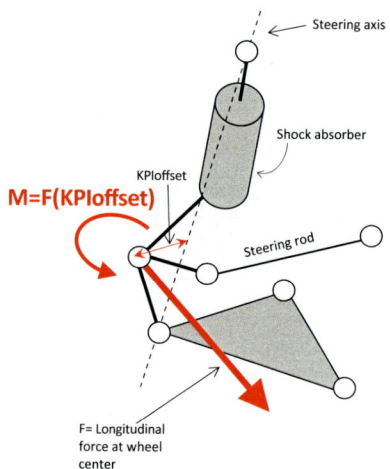

Figure 13 – Left side of the front suspension on which the longitudinal force due to the passage of the obstacle acts.

KPIoffset must similarly avoid having a large variability during the suspension operation to not generate a bad *steering feeling* in the traction phase. In fact, in traction, that is equivalent to two longitudinal forces at wheel center, symmetric and directed as in the following figure, when the *KPIoffset* left and right are the same or slightly different, the two reaction moments cancel each other and there are no vibrations on the steering wheel. If, on the other hand, during cornering motion, due to the bump travel on the external wheel and the rebound on the internal one, situations arise in which the *KPIoffset* differ too much, the two moments of reaction do not cancel each other out and produce a sudden and unwanted torque at the steering wheel. If in bump travel the *KPIoffset* increased and in rebound it decreased enough to diversify the two torques M in a non-negligible way, there would be a greater torque on the outer wheel which would tend to close the steering wheel towards the inside of the curve. It is appropriate that the increase of the *KPIoffset* in bump travel of the wheel also corresponds to the increase of the *caster trail* so that the torque increase due to the *caster trail* compensates for the tendency to "close" due to the *KPIoffset*.

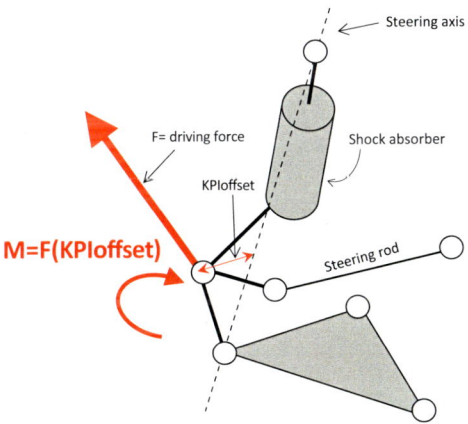

Figure 14 – Left side of the front suspension on which the longitudinal force due to traction acts.

A further consequence of a high *KPIoffset* lies in the excessive lateral displacement of the wheel center when steering, corresponding to a high pumping of the drive shaft which can cause the coupling to come out of the gearbox. Another effect not to be forgotten, on front-wheel drive sports cars, is the closing of the steering (steering torque towards the center of the curve) caused by the increase in traction force on the wheel outside the curve following the intervention of the self-locking differential, which notoriously shifts the drive torque and produces a sudden yaw of the vehicle towards the center of the trajectory.

From the previous observations it can be concluded that the objective is lowering as much as possible the *KPIoffset* to avoid annoying effects on the steering wheel. Generally, the decreasing in *KPIoffset* corresponds to the lowering of the KPI. In the Double wishbone of the figure 12 it can be seen that to lower the transverse inclination of the steering axis it is necessary to either move joint A inwards or move joint B outwards.

The limit to the displacement of joint A is the increase in *KPIoffset*, while the limit to the displacement of joint B is the minimum distance towards the rim.

An artifice to free oneself from the encumbrance towards the rim, widely used for high segment cars, is the replacement of one of the arms (or both) with either two links or connecting rods, so that the center of rotation becomes the point of intersection between the median lines of the links. In the following figure, the graphic representation of this replacement with the evident dependence of the inclination of the steering axis on the orientation of the two links is shown. Since point *B* is determined by the intersection of the extensions of the two links, there is no need to approach the rim or tire to lower the *KPI*. An architecture of this type, which takes the name of virtual centers suspension, is not easy to manage both in terms of design and industrial production because the lines of the two links must never be skewed, otherwise it becomes difficult to define and precisely control the position of point *B* with consequent complications on the trend of torque at the steering wheel.

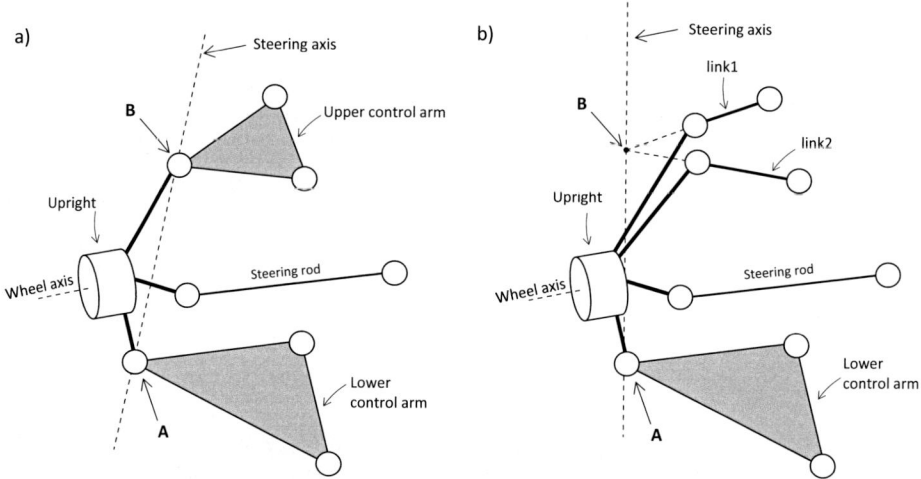

Figure 15 – Comparison between the inclination of the steering axis of an ordinary Double wishbone (a) and a Double wishbone with an upper virtual center (b).

The following figure shows the schematization of a McPherson suspension in which it is even more difficult to lower the transverse inclination of the steering axis. This is because of the shock absorber encumbrance towards the wheel, which does not allow point B to be moved towards the outside, and also for the *KPIoffset* containment which does not allow the displacement of point *A* towards the outside of the vehicle.

The steering system

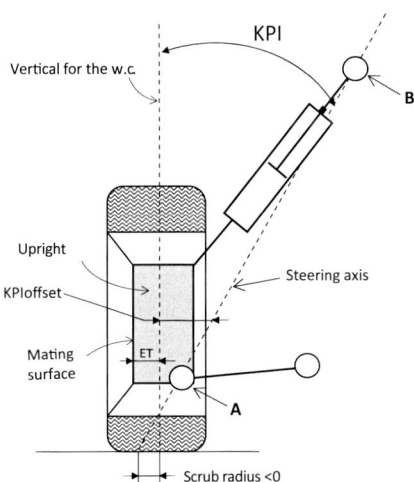

Figure 16 – McPherson showing the *KPIoffset* and the rim offset (ET).

From what has just been said, we can say that the McPherson architecture, due to the difficulty of lowering the *KPI*, guarantees less *camber* gain in steering compared to the Double wishbone. This is therefore more suitable where greater steering precision is sought, although it comes at a higher cost.

7.6. The scrub radius

With reference to the previous figure, the scrub radius is the distance in y, evaluated on the ground, between the trace of the steering axis and the trace of the vertical passing through the wheel center.

The scrub radius is a very important quantity for the torque at the steering wheel deriving from friction and therefore for the *steering feeling*. This is because it represents the distance between the longitudinal forces on the ground (resistance to advancing or braking) and the steering axis. It is positive when the trace of the steering axis is inside the vehicle with respect to the wheel center, as happens in the Double wishbone shown in figure 12, and it is negative when the trace is on the outside as for the McPherson of the previous figure.

It is important that the value of the scrub radius never changes its sign during the operation of the suspension (bump-rebound travel, steering, etc.) because if this happened it would change the sign of the reaction torques exerted by the longitudinal forces creating an annoying trend inversion on the *steering feeling* that would be perceived as a lack of torque at the steering wheel during some maneuvers. This can happen when scrub radius is small.

An important quantity for defining the scrub radius is the rim offset. This is the distance, along the axis of the wheel, between the mating surface of the rim on the disc and the wheel center. The latter is the intersection point between the wheel axis and the longitudinal symmetry plane of the tire. The rim offset, indicated by *ET* in the following figure, is positive when the wheel center is internal with respect to the mating surface on the disc.

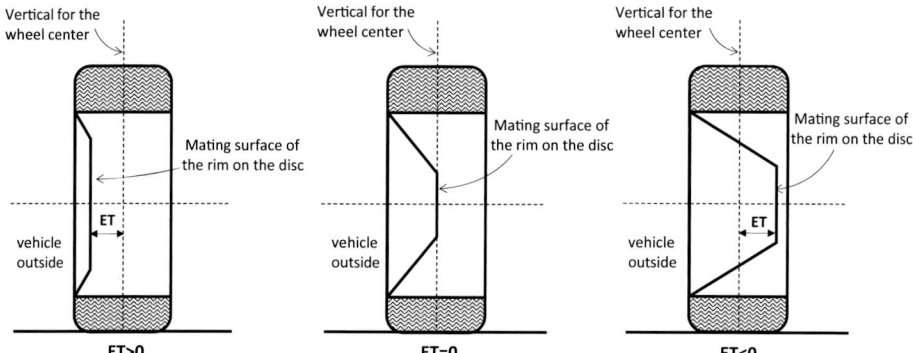

Figure 17 – Rims sectioned with a plane passing through the wheel axis. Positive rim offset (ET > 0): The wheel center is internal with respect to the mating surface of rim. Negative rim offset (ET < 0): The wheel center is outside with respect to the mating surface of rim.

The rim offset influences the value of the scrub radius and of the *KPIoffset* as seen in the following figure in which the assembly of two rims with *ET > 0* and *ET = 0* on the same McPherson type suspension was simulated. In particular with *ET > 0*, the scrub radius is negative while with *ET = 0* the scrub radius becomes positive, the *KPIoffset* increases, and the external width of the car is also increased by 2 times the rim offset > 0. The variation of these quantities influences the response to the steering wheel. For this reason, when buying a new rim it is advisable to make sure that the rim offset is identical otherwise annoying and inconsistent variations of the torque at the steering wheel can occur.

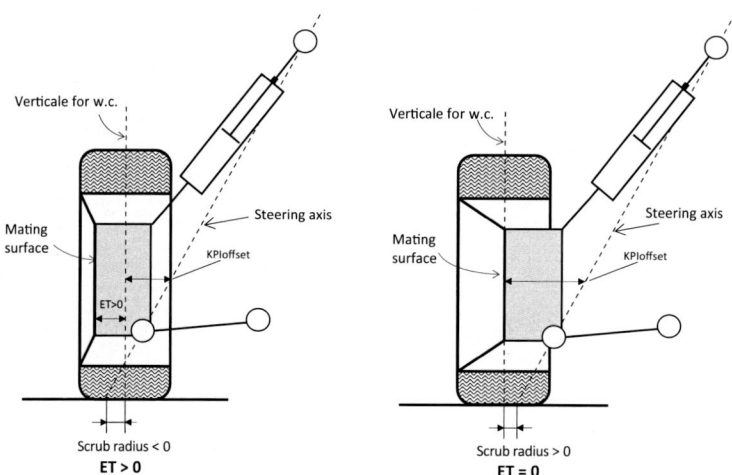

Figure 18 – Installation of wheels with different rim offset on the same suspension and consequent variation of the scrub radius.

The steering system

In production cars with different tire sizes, the scrub radius varies according to the size of the tires adopted. In order to use a wider tire and maintain the same width of the vehicle, it is necessary to increase the rim offset by half the difference in width between the tires.

The sign of the scrub radius depends on the purpose of the car and the suspension architecture used. In a McPherson suspension it is very difficult to set a scrub radius>0 without having an excessive value of the *KPloffset*, a problem that does not occur in Double wishbone or in suspensions with virtual centers that have more possibilities to move the steering axis. In general, sports cars, where possible, have scrub radius> 0 because this characteristic guarantees greater *feedback* on the road surface. In fact, in a different adherence condition between the left and right side of the vehicle, the car with positive scrub radius, in the event of braking, manifests a torque at the steering wheel and a yaw torque of the vehicle towards the area with greater adherence, returning to the user *feedback* on road conditions. Under the same adherence conditions, the vehicle with negative scrub radius does not provide any information on the road surface because the effect of the aforementioned yaw torque does not manifest itself.

We shall now explain the phenomenon through the following figure where a front axle with negative scrub radius is represented. Since there is greater adherence on the left side, the braking maneuver produces a longitudinal force *Rl* greater than *Rr* with the consequent tendency of the vehicle to yaw counterclockwise. The torques *Ml* and *Mr*, with *Ml>Mr*, due to the negative scrub radius and to the longitudinal forces, rise up the steering kinematics, producing a torque *C* on the steering wheel controversial to the previous yaw motion which guarantees the realignment of the vehicle. In this case, the perception of the yaw deriving from the different adherence conditions is greatly reduced because it is immediately counteracted by the torque at the steering wheel.

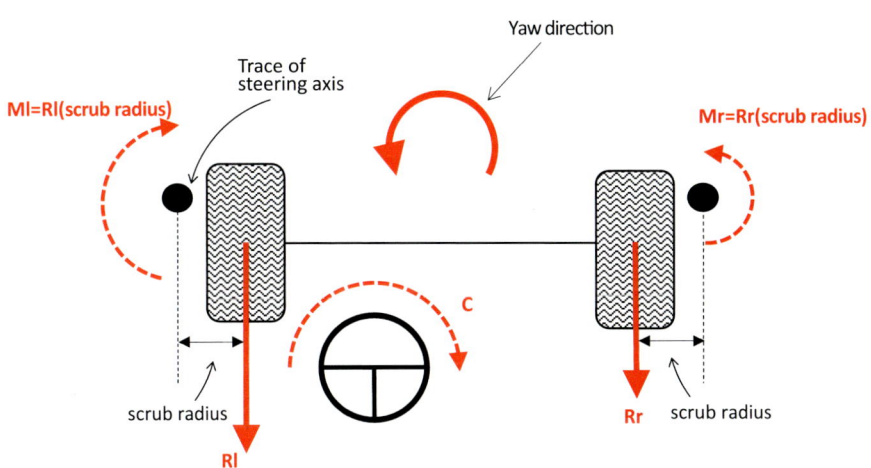

Figure 19 – Schematization of the braking loads in the case of negative scrub radius.

Automotive suspension

The situation is opposite in the case of a positive scrub radius because the moments *Ml* and *Mr* generate a torque *C* in the same direction as the yaw of the vehicle which emphasizes the phenomenon and forces the user to correct to stay on the trajectory. The latter behavior is less comfortable but is more suitable for sports cars because it returns information on the road surface to the user. The range of scrub radius values is approximately from -15 to +15 mm even if there are cars whose scrub radius is also 25-30 mm.

7.7. The steering of the wheels

The steering kinematics is the mechanism that determines the trend of the steering angles of the two wheels. The relationship between the internal and external steering angles greatly influences the lateral dynamic behavior of the vehicle because it determines the slip angles with which the wheels face the curve. The following figure shows, the ideal steering angles of *Ackermann* where the wheels turn so that their axles join on the axis of the rear wheels.

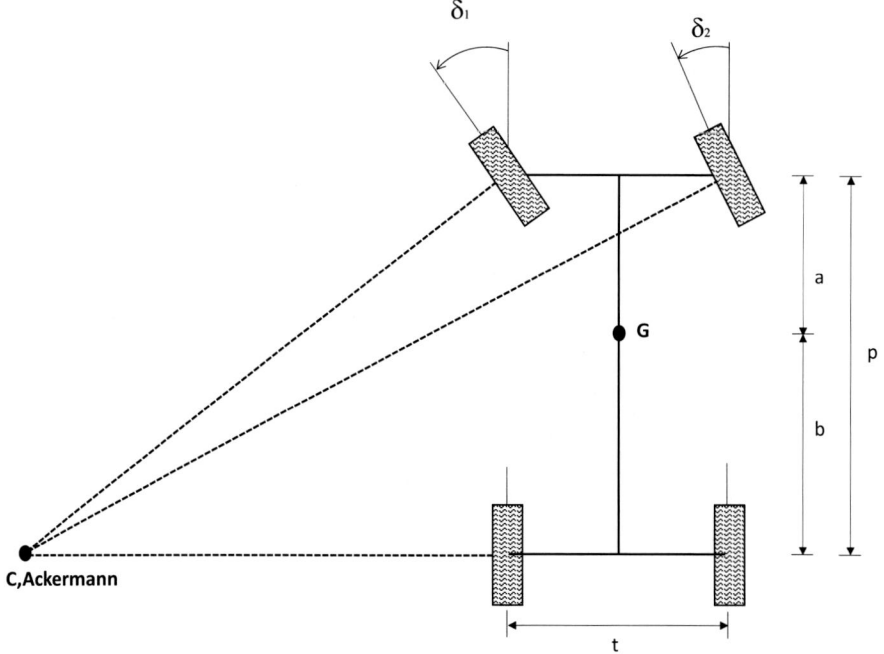

Figure 20 – *Ackermann* steering with instantaneous center of rotation $C_{,Ackermann}$.

The relationship between the wheel angles is as follows:

$$\cot \delta_2 - \cot \delta_1 = \frac{t}{p}$$

This is an ideal kinematic condition because it is the only one that, at low lateral acceleration, does not produce slip angles as the wheel centers move following the rolling of the tires. The lack of slip angles means the absence of lateral forces that strain tires and which are felt with increased effort at the steering wheel. Otherwise, when the steering does not follow formulation of *Ackermann*, slip angles emerge with zero lateral acceleration and therefore useless exploitation of the adherence capital already at low speed. These slip angles are necessary in order to the center of instantaneous rotation *I.C.* arises. This, contrary to the ideal case, is the result of an elastic deformation of the system since the line *a* does not meet the junction of the lines *b* and *c*, as shown in the following figure.

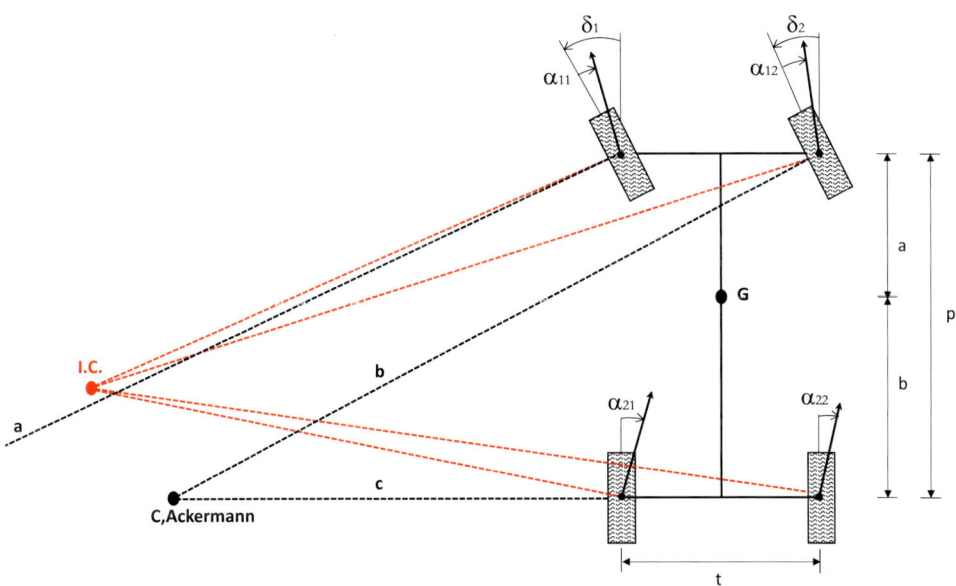

Figure 21 – No-kinematic steering with evidence of the slipe angles occurring even at low speed thanks to which the center I.C. different from the kinematic one is identified.

We shall now define the Ackerman percentage with the following formula:

$$\%Ackermann = \frac{\delta 1 - \delta 2}{\delta 1 - \delta est(ack)} \cdot 100$$

In which $\delta est(ack)$ is the value of the external wheel angle corresponding to the ideal *Ackermann* steering relative to the internal angle $\delta 1$. When the wheels steer parallel this percentage is zero. When the outer wheel turns more than the inner wheel the percentage is negative, and when the wheels follow Ackerman law the percentage is 100. We will call *pro-Ackermann* the positive steering percentage and *anti-Ackermann* the negative steering percentage. Once the internal angle of the curve is fixed, the *Ackermann* error function is defined, equal to the difference between the external wheel angle and the

Ackermann wheel angle. In road cars, this difference is almost always positive. The following figure shows the trend of the *Ackermann* error, the *Ackermann* percentage and the outer *Ackermann* angle as a function of the angle of the inner wheel in a left turn maneuver of a small car. The figure shows how the external angle is always greater than the ideal one, as happens in most vehicles. In addition, the *Ackermann* percentage increases with increasing steering to allow greater agility at low speeds.

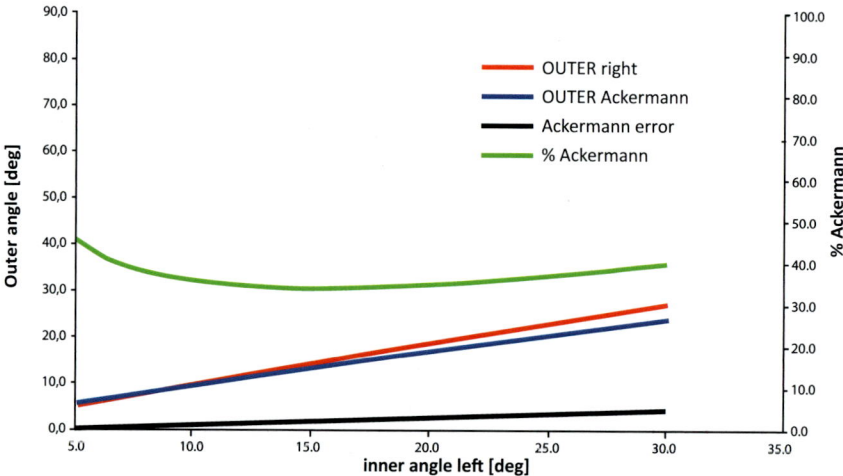

Figure 22 – Trend of external angle to the curve, *Ackermann* error and *Ackermann* percentage as a function of the inner angle, in a left turn maneuver.

In road cars, the *Ackermann* percentage remains low to induce less steering impulsiveness and the desired understeer which makes these vehicles more easily controllable for inexperienced users. In sports cars, which require rapid changes in direction, we tend to work with higher *Ackermann* percentages, bearing in mind that greater agility always implies less stability. For example, in cars with positive scrub radius, the self-steering induced when braking on asymmetrical road surfaces is certainly more amplified by a *pro-Ackermann* geometry. Another limitation of this trend lies in the fact that the greater steering angle of the inner wheel becomes dangerous when the *sliding limit* of the tire characteristic is lowered too much due to the decrease in vertical load. Generally, the characteristics of the front tires, in the presence of a load transfer, follow the trend of the figure below, in which the inner tire is closer to the crisis because as the load decreases the *sliding limit* is lowered (slip angle corresponding to the maximum force).

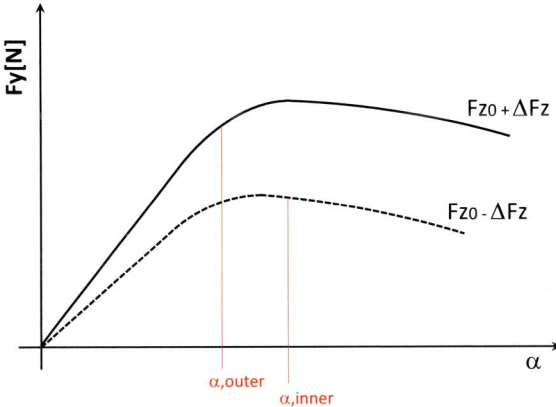

Figure 23 – Trend of lateral forces in the presence of a significant load transfer. In this case, the inner wheel works in the unstable section of its characteristic.

The *pro-Ackermann* geometry produces more slip on the inner wheel which is always more steered than the outer wheel. Therefore, the setting of the *Ackermann* percentage or the *Ackermann* error must take into account that, in the various operating maneuvers, the *α,inner* can exceed the *sliding limit* by generating the crisis. The following figure describes a different behavior in which the *sliding limit* of the inner tire, thanks to the work of the suspensions (*camber* gain and bump-steer), undergoes a shift to the right. This is a favorable situation for a *pro-Ackermann* geometry because it is possible to impose a greater steering angle on the inner wheel without sending the tire into crisis.

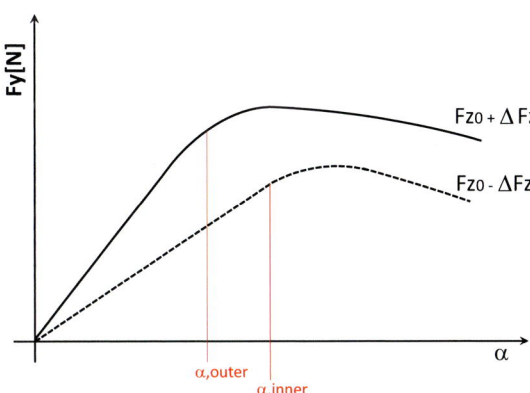

Figure 24 – Trend of lateral forces in the presence of a significant load transfer. In this case, thanks to suspension work, the inner wheel still works in the stable section of its characteristic.

In general, there is a tendency to adopt *pro-Ackermann* geometries when the *sliding limit* of the inner wheel is not lowered too much and *anti-Ackermann* for the opposite situation. The adoption of an important *pro-Ackermann* trend requires complex analyses which consist of evaluating the characteristics of the tires based on the load transfers and the wheel angles assumed during the various maneuvers. To give a qualitative motivation for these behaviors, it should be noted that the tire inside the curve always works with *camber* and *toe* angles opposite to the ones outside the curve. In particular, as shown in the following figure, the characteristic of the inner tire, which translates rightwards due to the *camber*, produces an increase in the *sliding limit*.

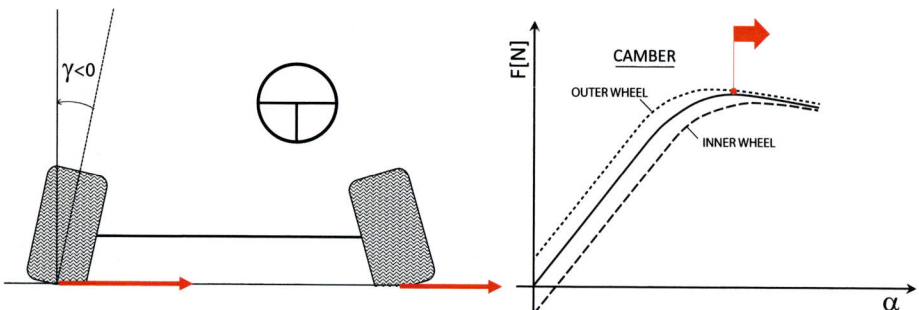

Figure 25 – Qualitative representation of the rightwards translation of the inner tire characteristic of the curve due to *camber*. In the front view of the axle, the lateral forces emerging when cornering are represented.

On the contrary, the static *toe-out* moves the characteristic of the inner tire leftwards by lowering the *sliding limit*, as shown below. Based on these qualitative observations we can assert that, as the *camber* gain in bump travel (compression of the spring) increases, which corresponds to the decrease of the loss in roll for the two wheels (see chapter 6), the *sliding limit* of the internal wheel decreases. Instead, as the divergence in bump – travel (compression of spring) increases, in the case of a symmetrical kinematic curve, the *sliding limit* increases.

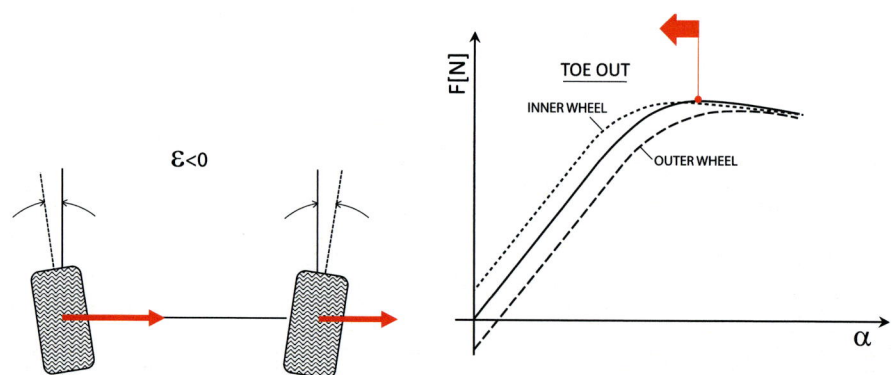

Figure 26 – Qualitative representation of the leftwards translation of the inner tire characteristic caused by front *toe-out*. In the plan view of the axle, the lateral forces emerging when cornering are represented.

On some rear-engine cars, with little weight on the front axle, it is very easy to find an excessive lowering of the load on the inner tire, which always corresponds to the leftwards shift of the *sliding limit*. In these situations there is a tendency to adopt a steering geometry that is only slightly *pro-Ackermann* and in extreme cases *anti-Ackermann* (external steer angle greater than internal). Another reason for an *anti-Ackermann* choice lies in the fact that the internal wheel that is too steered creates greater resistance to advancement with a consequent increase in the temperature of the tire. In fact, in some set-ups of very fast sports cars, *anti-Ackermann* geometry is preferred in order not to lose speed. On the other hand, the braking of the inner wheel, called *drag force*, positively affects the sudden changes of direction because the associated longitudinal force produces a yawing torque. For this reason, the *pro-Ackermann* steering becomes effective in not very fast paths at low radii of curvature. However, this must not be exaggerated too much because the *drag force* trend in relation to the slip angle is similar to that of the lateral force as it decreases after exceeding the value of the *slip angle* corresponding to the peak. This implies that, after the peak, the internal *drag force*, which initially produced a yawing torque, tends to decrease its oversteer effect and the understeer effect due to the external *drag force*, which is still in its increasing phase, prevails. For this reason, when deciding to adopt *pro-Ackermann* geometries, it is important to also consider the internal *drag force* trend.

On very fast cars, where the aerodynamics that produces downforce such that the vertical load transfer becomes negligible, it is possible to adopt large *Ackermann* percentages, which in some cases exceed 100% (internal wheel that steers at a greater angle than *Ackermann*).

Unfortunately, there is no defined procedure to identify the right Ackerman percentage. This is because it is also related to the slip angles, variables dependent on the lateral acceleration. However, once the trend is set, the right percentage is defined experimentally or by calculation.

In summary, the trends applied to wheel steering are described below.

Low percentage *Ackermann* or *Anti-Ackermann* steering:
- Cars in which the inner tire undergoes a significant lowering of the *sliding limit*;
- Rear-engine cars with little weight on the front axle;
- Very fast sports cars so as not to lose speed due to the internal *drag force.*

Pro-Ackermann steering:
- Cars for fast paths with low radii of curvature which require a certain agility in direction changes. The percentage increases as the weight on the front axle increases.
- Cars with large aerodynamic downforce for which the loss of load on the inner tire becomes negligible.
- Cars in which the inner tire does not undergo a significant lowering of the sliding limit thanks to the elastokinematics of the suspensions.

7.8. The steering mechanism

Ackermann steering is difficult to achieve, the only steering mechanism capable of guaranteeing ideal steering is the Bourlet one but it is not used because its management is complex both from the point of view of maintenance and reliability. Other kinematics that approximate the ideal steering are the Janteaud kinematics and the Panard kinematics, shown in the following figure, but they are similarly of very limited use.

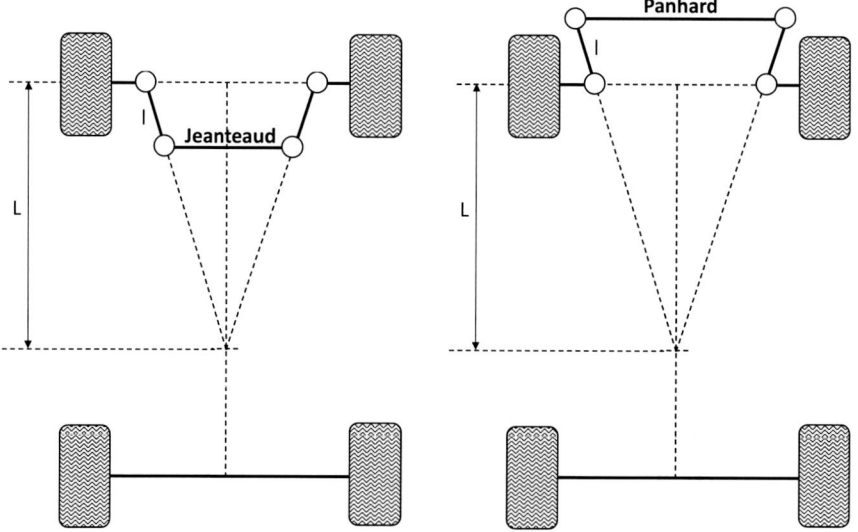

Figure 27 – Janteaud and Panhard mechanisms.

A feature common to all steering kinematics lies in the fact that in order to approach the ideal steering, the junction of the extension of the levers *l* is inside the vehicle and the lower the distance L is, the more the *Ackermann%* increases. It is therefore clear that in kinematics similar to the Panhard ones, with the steering box advanced with respect to the center of the wheel, the *pro-Ackermann* configuration is more difficult to achieve due to the risk of interference between the steering levers and the rim.

The most commonly used steering mechanism in vehicles equipped with suspension is the one shown in red in the following figure.

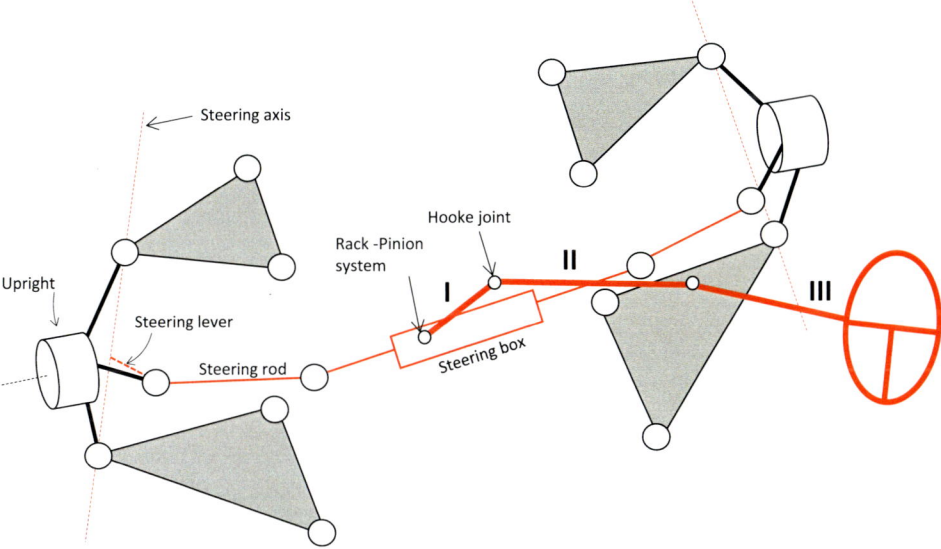

Figure 28 – Steering kinematics normally used on Double Wishbone and McPherson front suspension.

It consists of:
- The steering column consists of three sections connected to each other (in *P2* e *P3* of figure 30) by Hooke or cardan joints.

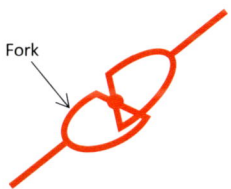

Figure 29 – Hooke joint.

The relative inclination of the three sections determines the level of homokinetics of the steering system which defines the difference between the angle imposed on the steering wheel and the angle measured at the pinion. This difference varies during the rotation of the steering wheel and produces a variable steering ratio with the rack travel. The ideal homokinetic situation, which assures a constant steering ratio, is fully guaranteed when:
1. The angle between the plane that contains axis *I* and axis *II* and the plane that contains axis *II* and axis *III* must be equal to the phasing angle of the forks (angle between the planes of the forks) of section *II*.
2. The solid angle between axis *I* and axis *II* and the solid angle between axis *II* and axis *III* must be equal.

Automotive suspension

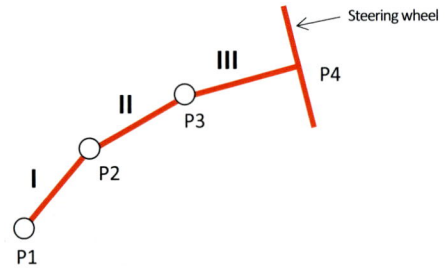

Figure 30 – Side view of the three sections of the steering column.

According to what has been said, if the sections *I*, *II* and *III* lie on the same plane, the angles *I-II* and *II-III* are the same and the relative angle between the forks of section *II* is zero, the steering line is homokinetic. From a practical point of view this means that the following conditions are homokinetic:

    a) Axes *I-II-III* coplanar, forks of the intermediate section coplanar and equal angle between sections *I-II* and *II-III*, as in the following figure.

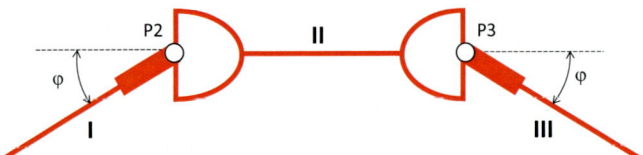

Figure 31 – Homokinetic condition with non-parallel input and output axes.

    b) Axes *I* and *III* parallel (and therefore coplanar), coplanar intermediate sections forks and equal solid angle between sections *I-II* and *II-III*, as in the following figure.

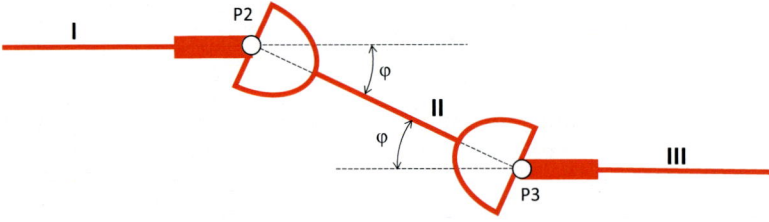

Figure 32 – Projection of the steering mechanism on the plane containing I and III. Homokinetic condition with parallel input and output axes.

The following figure shows, the variability of the relationship between the steering wheel angle and the pinion angle as a function of the rack travel, in the case of a non-homokinetic steering column. On the other hand, when the steering column is homokinetic, this ratio is always 1. It is quite intuitive that in sports cars a non-homokinetic behavior is to be avoided, as it produces the undesired variability of the steering ratio which is both a source of delays and of non-uniform response of the vehicle.

The steering system

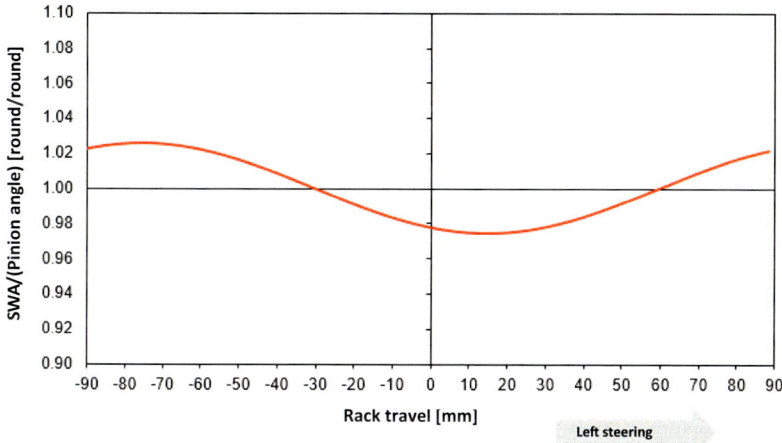

Figure 33 – Trend of the (steering wheel angle)/(pinion angle) ratio as a function of the rack travel in the case of a non-homokinetic steering column.

- A central steering box with rack and pinion system whose transmission ratio contributes to the steering ratio. It is placed in a forward position with respect to the wheel center when the vehicle is a sports car and a lower position of the steering wheel is desired in order to lower the center of gravity of the driver. In fact, in relation to part *a)* of the following figure, which shows the rack housing set back with respect to the wheel center, if it was desirable to lower the steering wheel the angle between sections *II* and *III* will have to be increased beyond the limit allowed by the mechanism, as section *I* cannot be moved due to the encumbrance towards the pedal board. This does not happen in the case of the rack housing advanced with respect to the wheel center, because the section *I* has fewer constraints and because the greater length of the column allows for smaller relative angles with the same lowering of the steering wheel.

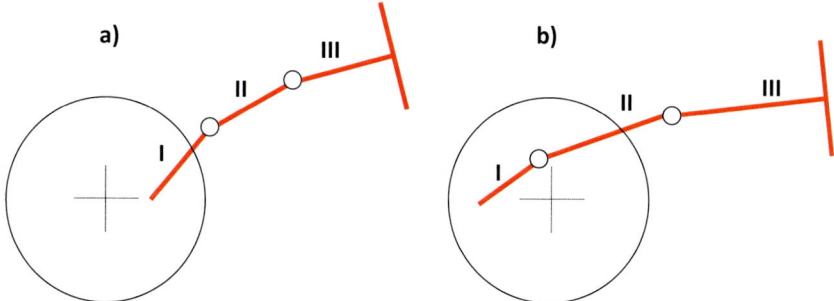

Figure 34 – Side schematic view of the steering column in the case of rack housing set back with respect to the wheel center (a) and rack housing advanced with respect to the wheel center (b).

- The steering rod (left and right) that connects to the upright with a ball joint. It pushes the upright in rotation around the steering axis and regulates the *bump steer*.

When, for example, the steering box is set back and the extreme rack housing side of the steering rod is higher than the upright side, there is a tendency to close in bump travel. Below the schematic representation of the front view of the left side of a Double wishbone is shown. It is evident that the external point of the steering rod moves following a circular trajectory with lateral displacement towards the outside (wheel closure) in case of bump travel and towards the inside (wheel opening) in case of rebound travel.

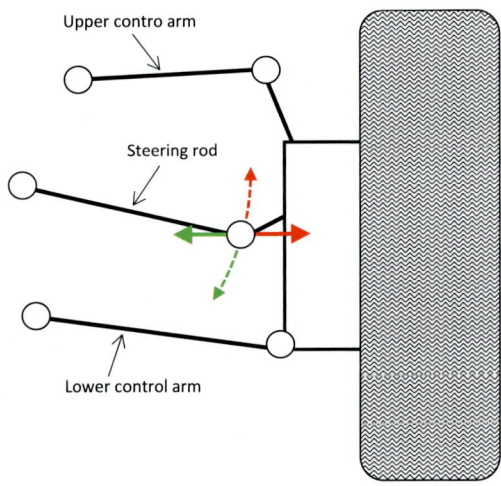

Figure 35 – Front view of a Double wishbone left side with the extreme rack housing side of the steering rod higher than the extreme wheel side.

Obviously, this is only a tendency that affects the relationship between *camber* gain and *bump steer* due to the inclination of the instantaneous rotation axis previously treated (chapter 6). In fact, in the suspension setting phase it is appropriate, if possible, to start from the configuration in which the straight line of the steering rod intersects the axis of instantaneous rotation and vary the height of one of the two extremes in order to obtain the desired *bump steer*.

· The steering levers, which represent the distances between the end of the steering rod (ball joint center on the upright) and the steering axis, have an inclination that is not that of the physical arm on the upright but is that of the straight line constructed as just described. Since the length of the steering lever defines the steering ratio, a very common technique to make the car more direct is to lower this length, but this has the collateral effects of excessive bump steer and the increase of steering wheel torque.

8. VEHICLE DYNAMICS
Performance evaluation

8.1. The evaluation tools

The tools used to evaluate the dynamics of a vehicle are different depending on the stage in which the car lies. In the design phase it is necessary to adopt a virtual-numerical approach which consists in the use of calculation codes based on the mechanics of Euler-Lagrange also called *multibody* technique. *Multibody* codes allow to create a mathematical model of the vehicle capable of simulating the different operating maneuvers and to evaluate their dynamic behavior. Evaluating the dynamic behavior of a system means knowing the details of the forces and the consequent displacements on each component. It also means identifying the most sensitive parameters and acting on them to modify the overall dynamic behaviour. *Multibody* codes are used to simulate the behavior of subsystems such as suspension, steering, etc. and to simulate the behavior of the entire vehicle. They allow us to visualize the simulation of the maneuver and to get a first impression of the operation. Despite this, they are quite complex tools because they contain a large amount of information to keep under control and not all of them are immediately visible. Generally, by using these codes, it is possible to define the dynamic simulation of the entire vehicle by first passing through the various subsystems.

In particular, starting from the initial geometry of suspension and steering, the final positions of the characteristic points and the stiffnesses of the bushing are identified by simulating the different operating maneuvers. Subsequently, having defined the optimal geometry of the suspensions, they are assembled on a rigid body with the correct arrangement of the masses and, after having assembled the tires with the correct characteristics, the appropriate handling maneuvers are carried out to evaluate the dynamic behavior of the virtual vehicle. The numerical modeling of tires involves the use of the *Pacejka* formulation which can be imagined as a black box that provides the slip angle having as input the characteristic angles (*camber*, *toe* etc.) and the forces exchanged between tire and road. Each compound and size of the tire has its own specific formulation, and in order to correctly interpret the vehicle dynamics it is important to equip the *multibody* model with the right tire.

We describe below two of the most important maneuvers for the handling evaluation:
1. Steering pad at constant radius. The vehicle makes a constant radius curve at increasing speed.
2. Step steer. As the vehicle travels straight at constant speed, a certain steering wheel

angle is imposed in a relatively short time.

These maneuvers, which provide useful information on the dynamics of the vehicle, are carried out both in a virtual environment, using the calculation codes, and on the real vehicle. The execution of the virtual tests has the purpose of predicting and improving the behavior of a vehicle. On the other hand, the execution of the tests on the finished vehicle has the purpose of validating what is foreseen in the design phase and to approve the vehicle because it complies with the project specifications. Generally, commercially available *multibody* computing codes, if used correctly in terms of modeling and of input, are able to reproduce the behavior of a dynamic system with a very high level of precision. This allows us to evaluate and validate different design solutions without the need to create any physical prototype, guaranteeing considerable savings on development costs.

The carrying out of the aforementioned maneuvers on the real vehicle requires the measurement of the following quantities:

- Steering wheel angle: it is measured with a rotational transducer applied to the column and through the steering ratio and the knowledge of the steering kinematics the wheel angles are calculated.
- Accelerations in the three directions: they are measured with capacitive type accelerometers positioned on the center of gravity of the vehicle. The distance of the center of gravity from the front axle is easily calculated by knowing the distribution of the masses. As for the height, we start with the height from the ground of the driver's hip and then correct it later as we will show.
- Yaw rate, pitch and roll: measured with a gyroscope applied near the center of gravity of the vehicle. Generally, the position of the gyroscope coincides with that of the accelerometers because they are anchored to the same structure.
- Wheel displacements in bump and rebound travel: obtained by multiplying the displacement of the shock absorbers, measured by linear transducers, by the known lambda ratio.

8.2. Steering pad

In the Steering pad with constant radius, the vehicle travels on a circular path at increasing speed until the crisis is reached. The speed increases very slowly to allow the vehicle to stabilize at successive lateral acceleration values.

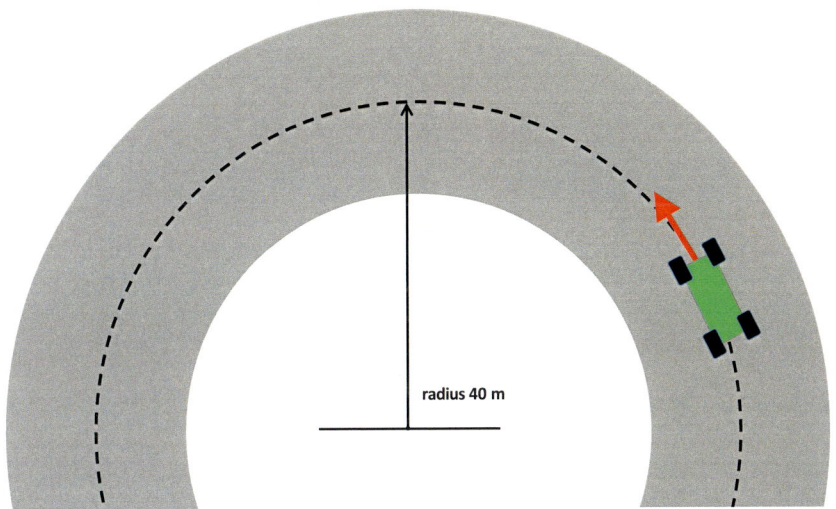

radius 40 m

Figure 1 – Steering Pad maneuver with radius of 40 m.

This maneuver is used to evaluate the dynamic behavior in stationary conditions, that is when the vehicle stabilizes at the different accelerations reached, without taking into account the transient. In these conditions, the objective quantities are measured to construct the understeer curve, the sideslip curve, the roll gradient and the torque at the steering wheel.

8.2.1. The understeer curve

Having the steering wheel angle and the lateral acceleration of the center of gravity available, we can construct the understeer curve, which, as mentioned, is the trend of the steering wheel angle as a function of lateral acceleration in subsequent stabilized conditions, without taking into account the transient to achieve that acceleration.
The following figure shows the understeer curve of a generic car in which the lateral acceleration is expressed in multiples of *g*. It consists of a linear section until *0.55g* and a non-linear section that goes from *0.55g* to the maximum acceleration (*AYMAX*) reachable by the vehicle which, in this case, is approximately *0.92g*. The first section has a constant slope because, referring to the equivalent bicycle for simplicity, the axles are working in the linear section of their characteristic curve. Therefore, to increases in lateral acceleration correspond linear increases of the front and rear slip angles, according to equations 4.8. which we rewrite for convenience

Automotive suspension

$$\alpha_f \; [radians] = \frac{M_f A_{Gy}}{C_{\alpha,f}}$$

$$\alpha_r \; [radians] = \frac{M_r A_{Gy}}{C_{\alpha,r}}$$

As long as the axles have a linear trend, which means constant slip stiffness $C\alpha$, the ratio between acceleration increase and slip increase is constant and is $Mf/C\alpha,f$ for the front axle and $Mr/C\alpha,r$ for the rear one. Consequently, as evident in equation 4.13, if these ratios remain constant, the trend of the steering wheel angle is linear and the slope in the origin Kus is the understeer gradient which in this case is 28 [deg/g].

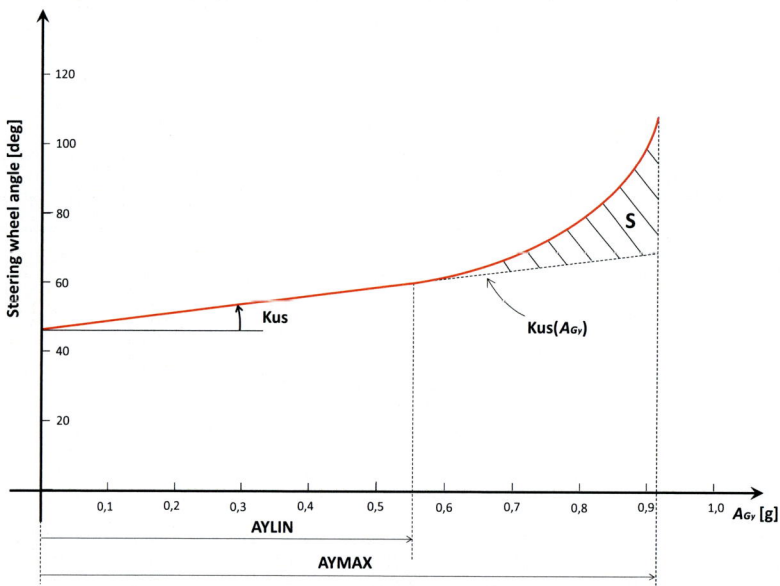

Figure 2 – Understeer curve in a constant radius steering pad maneuver.

The linear section ends when at least one of the axles assumes a non-linear behavior. This occurs when the corresponding tires are in that section of the curve in which the ratio between the increase in force and the increase in the slip angle decreases as the force increases, which corresponds to section *II* of the axle characteristic in the following figure.

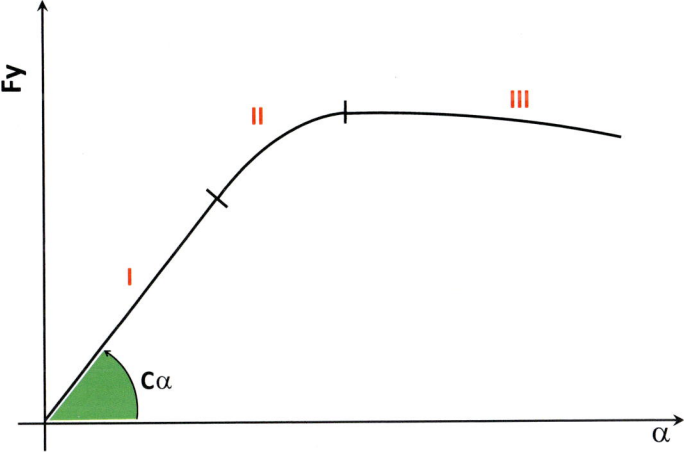

Figure 3 – Axle characteristic example.

Notably, when the slope of the understeer curve increases, it means that the front axle has passed through section *II* because if, with the same increase in acceleration and therefore in lateral force, the increase in the front slip angle grows, the increase of the steering wheel angle also grows. Otherwise, if the rear axle enters in section *II*, the steering angle decreases and consequently also the slope of the understeer curve. It absolutely must not happen that the rear axle reaches section *II* before the front axle, otherwise the vehicle oversteers. This is a condition that hardly occurs in front-engine vehicles because the unbalance of the forward weight guarantees greater slip angle on the front axle. Excessive displacement of the weight distribution towards the front axle, however, limits the correct use of the axles while a more balanced weight distribution (50%), since it delays the transition to the non-linear section of the front axle characteristic, makes the understeer curve straighter. From what has been said, we can say that the length of the linear section of the understeer curve, identified in the figure by the *AYLIN* value, is proportional to the length of section *I* of the front axle characteristic.

Another characteristic of the understeer curve is the value of the area of the dashed surface *S* which represents the measure of the increase in the steering wheel angle when the two axles enter sections *II* and *III*. Area *S* is the difference between the area under the understeer curve and the area under the straight line $K_{us}A_{gy}$. If too high, the excessive difference in the increase in the steering wheel angle that the user feels when exiting the linear section of the curve could be interpreted as a lack of reactivity. With reference to the front axle, we can say that the more the curvature of section *II* increases, the more the area *S* increases. Furthermore, the slope of the understeer curve, near maximum acceleration, tends to increase if the front axle anticipates entry in section *II* of the characteristic. Instead, it tends to decrease if the rear axle anticipates the entry into the non-linear section. This is because if the increase in rear slip grows, the increase in the steering wheel angle decreases and consequently the slope decreases.

The desired condition is the one in which the understeer curve starts from a constant slope trend and undergoes a progressive increase that is downsized when the rear axle also enters its characteristic section *II*. If the rear axle never enters the trait II of the characteristic, the slope of the understeer curve increases too much and becomes infinite (vertical tangent) when the front axle reaches the *sliding limit* (slip angle relative to the maximum of the axle characteristic) and becomes negative when the axle enters the section *III*. This is also a condition to avoid because it leads to excessive steering wheel angles in the final area of the curve, to unwanted counter-steering when the slope becomes negative and to a decrease in the *AYMAX* because the rear tires are not used. This situation can also occur when the sizing of the rear suspension does not guarantee large sideslip angles.

We shall now make a comparison between the understeer curves in the following figure of two identical cars that differ only in the front suspension.

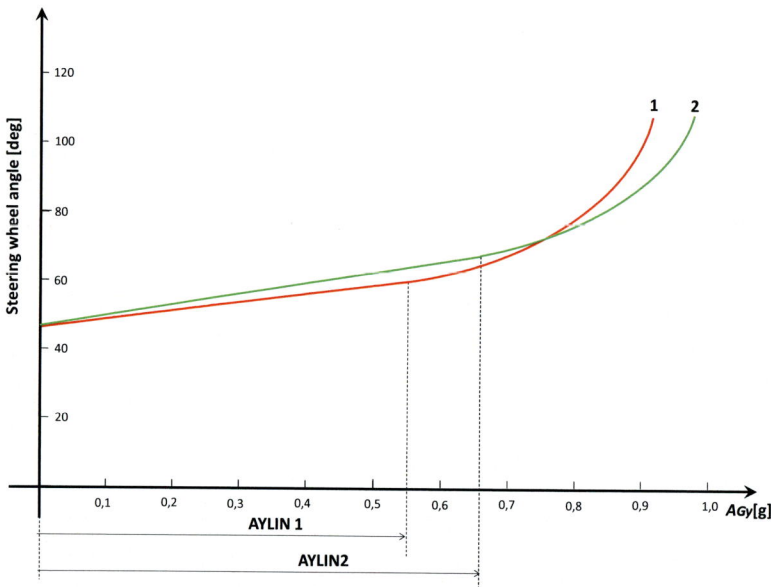

Figure 4 – Constant radius understeer curves of the same car equipped with two different front suspensions. 1 McPherson, 2 Double wishbone.

Car 1 is equipped with a McPherson suspension, car 2 is equipped with a Double wishbone suspension. Car 1 has the shorter linear section due to the fact that the McPherson has a non-linear *camber* gain which becomes a loss of *camber* after a certain value of bump travel (following figure) and this gives the axle a lower slip stiffness than the Double wishbone. This difference in performance can also emerge due to the bump suspension caused by the aerodynamic downforce, therefore it is advisable to consider this possibility in the presence of non-linear of the elastokinematic curves. In addition, car 2, thanks to the lower loss of *camber* under lateral load, due to the presence of two transverse arms, has a higher maximum acceleration and the lower *S* area.

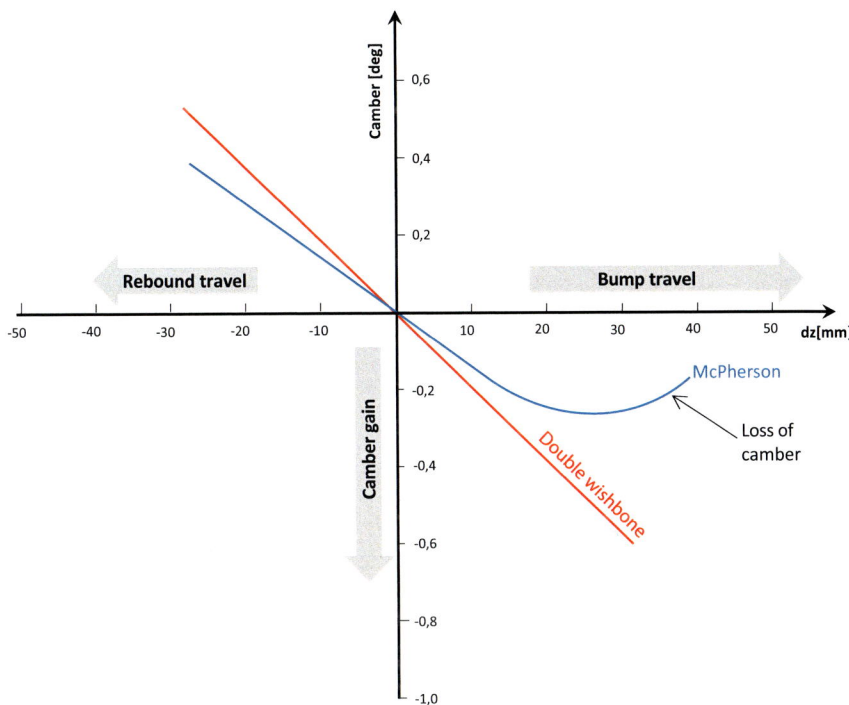

Figure 5 – Comparison between the variation of *camber* in parallel wheel travel of a Double wishbone and a Mcpherson with evidence of the McPherson reverse trend.

Another difference is in the *Kus* value which is greater for the Double wishbone despite the two cars having the same steering ratio. The reason lies in the fact that the Double wishbone has been designed with a greater wheel opening under lateral load, which increases the steering angle when cornering and delays the entering in the non-linear section of the axle characteristic allowing a higher *AYLIN* value.

We can conclude that car 2 is definitely more performing because it has the most progressive understeer curve and greater *AYMAX*.

The constant radius understeer curve is influenced by the elastokinematics of the suspension, by the characteristics of the tires, by the springs and by everything that influences the slip angles of the axles in the stabilized condition and not in the transient condition. It is not affected by shock absorbers and damping in general because they do not generate any force in stationary.

8.2.2. The sideslip angle

From expression 4.10, which we rewrite for convenience, it is clear that the sideslip angle depends only on the slip angle of the rear axle

$$\beta = \frac{b}{R} - \alpha_{,r} \quad (4.10)$$

The effect of the rear axle can be seen in the following figure showing the *AYLIN BETA* value which is different from the *AYLIN* of the understeer curve.

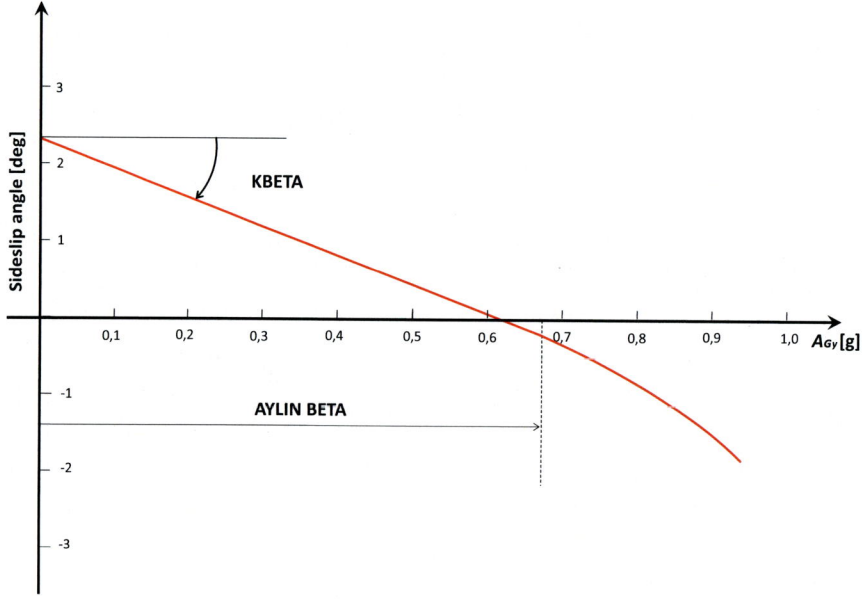

Figure 6 – Sideslip angle trend in the Steering Pad maneuver.

The *AYLIN BETA* value depends only on the rear axle, that is, on when the latter enters section *II* of its characteristic. As already mentioned, *AYLIN BETA* must be greater than *AYLIN* because, so that there is no slope reversal of the understeer curve, the front axle must enter in section *II* at a lower acceleration than the rear axle. If this does not happen, for acceleration greater than *AYLIN BETA*, the slope of the understeer curve decreases and the vehicle tends to oversteer.

The value in the origin is the kinematic sideslip angle, while the slope in the origin is the sideslip gradient which depends only on the rear axle, as evident from the already described analytical expression.

$$KBETA = \frac{M_r}{C_{\alpha,r}}$$

8.2.3. The roll gradient

The variation of the roll angle, as a function of lateral acceleration, mainly depends on the centrifugal force, on its distance from the roll axis and on the stiffness in opposite wheel travel of the suspensions. Since the force varies linearly with the acceleration and all the other quantities remain fairly constant, we can say that the roll angle trend is linear, as shown in the following figure. The slope of the aforementioned trend is the roll gradient which, depending on the type of car, varies from 1 to 7 *deg/g*. The roll angle trend is not linear when an increase in stiffness occurs on one of the two suspensions due to the intervention of the bump stop. This behavior is not appreciated because it involves a sudden change in the slope of the understeer curve, but if inevitable, it is preferable that the front bump stop intervenes in order to increase the understeer. Otherwise, if the rear bump stop intervenes, the car tends to oversteer.

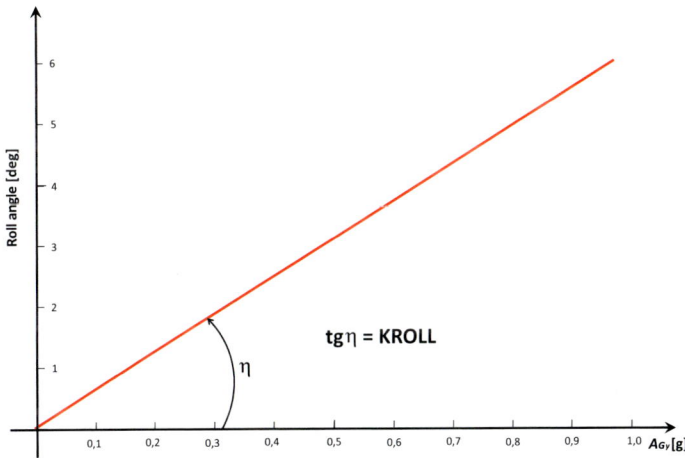

Figure 7 – Trend of the roll angle as a function of lateral acceleration.

The experimental measurements of the roll angle and lateral acceleration allow to evaluate the height of the center of gravity using the following relationship which has been also described in chapter 3:

$$\psi = \frac{M_s A_{Gy}}{Ka + Kp} d$$

From this relation, we can derive the distance *d* of the center of gravity from the roll axis and consequently, its height from the ground. Obviously, the front and rear roll stiffnesses and the position of the roll axis must be known. Roll stiffnesses are readily available both in the design phase and in the finished vehicle; the same applies to the position of the roll axis which is certainly known in the design phase and can be obtained from the measurement of the track variations in the finished vehicle, such as described in chapter 6.

8.3. The response of the vehicle in stationary

The modulation of the understeer curve changes the response of the vehicle in stabilized conditions making it more or less sporty. Below are the characteristic quantities of the curve and the parameters on which to act:

Kus
If it is low, the car is more direct. To lower the *Kus* is necessary:
On the front axle:
- Increase the steering ratio;
- Increase the static *camber*;
- Increase the *camber* gain in bump travel;
- Increase the *camber* gain in steering (*caster*);
- Decrease the loss of *camber* under lateral load;
- Adjust the static *toe-in* (closed wheels);
- Decrease the *toe-out* (opened wheels) variation in bump travel;
- Decrease the *toe-out* (opened wheels) variation under lateral load;
- Increase the *Ackermann* percentage.

On the rear axle:
- Decrease the static *camber* with respect the front axle;
- Decrease the *camber* gain in bump travel keeping it always greater than the front;
- Adjust the static *toe-out* (opened wheels);
- Decrease the *toe-in* (closed wheels) variation in bump travel;
- Decrease the *toe-in* (closed wheels) under lateral load.

AYLIN
If *AYLIN* increases, the operating range increases in which the delta steering angle is proportional to the delta acceleration which makes the car behavior more predictable and controllable. Cars with a higher percentage of weight on the front axle have lower *AYLIN* values because they undergo higher transverse forces that anticipate the entry into the non-linear section of the axle characteristic. The balance of the weight of the axles therefore helps to make better use of the tires because it improves the distribution of transverse loads. To increase *AYLIN* it is necessary to:
On the front axle:
- Increase the linear section of the axle characteristic by means of tires with the initial section *I* most extended (see chapter 2).
- Use suspension schemes that produce linear trends of the elastokinematic quantities with particular attention to *camber* and *toe* both in opposite wheel travel and under lateral load.
- When sizing the *camber* trend (gain and loss under lateral load) to decrease the *Kus* objective it required to consider the possible lowering of the *sliding limit* of the axle characteristic (see chapter 2).

- Increase, within the allowed limits, the wheel opening in bump travel and under lateral load to delay the entering in the non-linear section of the characteristic of the axle. The consequent increase in *Kus*, deriving from this action, can be remedied by intervening on the steering ratio.
- Increase the wheel track. The wheel track increase, which must always be modulated according to the rear one, lowers the load transfer and therefore the slip angle corresponding to the axle. With the same transverse force, an axle with a wider wheel track delays the entry into the non-linear area of the characteristic.
- Delay the intervention of the bump stops.

On the rear axle:
- Adjust the *camber* gain and loss under lateral load to lower the linear section of the characteristic of the rear axle so that the increase in sideslip angle limits the increase in the steering wheel angle.
- Adjust *toe-in* in bump travel and under lateral load to lower the linear section of the characteristic of the rear axle so that the increase in sideslip angle limits the increase in the steering wheel angle.
- Increase static *camber*, taking into account the front value, to lower the linear section of the characteristic of the rear axle.
- Set static *toe-in* to lower the linear section of the characteristic of the rear axle.

Remember that lowering the rear axle *sliding limit* must always ensure that AYLIN BETA is greater than AYLIN.

AREA S

The objective generally pursued is to have the area *S* not too large. This is because if it increases, a feeling of void response from the steering wheel is perceived in the last section of the understeer curve. If, on the other hand, the *S* area is too low, near the maximum lateral acceleration, the increase in the steering wheel angle does not undergo large variations causing the lack of perception of reaching the limit. To reduce the *S* area, it is necessary to increase the linear section on the front axle and lower it on the rear one, bearing in mind that the rear axle must not exit the linear section before the front one. As already explained above, the shape and value of area *S* depend on how the instants in which the two axles enter section *II* of the characteristic curve and the instant in which the front axle enters section III are arranged over time.

The action of the front bump stops, which must be sized to intervene before the rear ones and at higher lateral acceleration than *ALIN*, produces greater load transfer on the front axle with the consequent growth of the *S* area. It shall be reminded that is preferable bump stops do not intervene when cornering. Unfortunately, this is not always possible, especially for road cars, to identify the right compromise between bump stop clearance and stiffness in opposite wheel travel that allows it.

Automotive suspension

8.4. Step steer at constant speed

The step steer is performed by launching the vehicle at a certain speed and applying, in a very short time, a certain steering wheel angle which remains constant until the vehicle stabilizes. The sudden steering causes a sudden change in the dynamic quantities, which stabilize in the following interval in which the steering wheel angle remains fixed. This maneuver is used to evaluate both the dynamic behavior in the transient and that in stationary conditions. In the figure, the visual evidence of the effect of the same step steer on three different vehicles or with three different suspension set-ups is shown. After clarifying the dynamics of the step steer we will explain the objective differences between the cars.

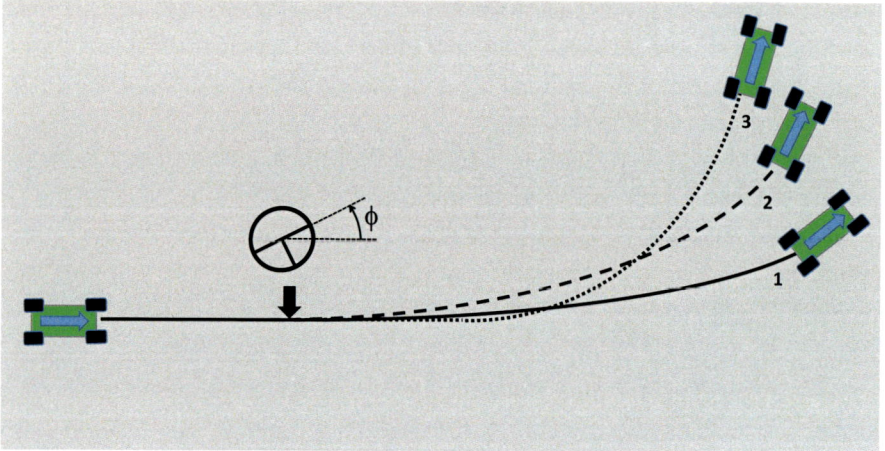

Figure 8 – Three different behaviors in a step steer maneuver.

The following figure shows the yaw speed and steering wheel angle trend in a step steer of 30 degrees at 100 km/h. It shall be recalled that the yaw speed (also called yaw rate) is the rotation speed around the vertical axis of the plane motion of the vehicle with respect to the road. It consists of two contributions: the first is the rotation of the vehicle center of gravity around the center of curvature, the second is the rotation of the vehicle around the center of gravity (rotation on itself). When the first contribution increases, at a constant longitudinal speed, the radius of curvature decreases, with the following approximate formula valid in stabilized, for small angles and not in the transient.

$$r = \frac{V_x}{R}$$

The trend of yaw rate *(r)* in the step steer shows a peak, an oscillation and a subsequent stabilized section whose characteristics depend on the performance of the suspension.

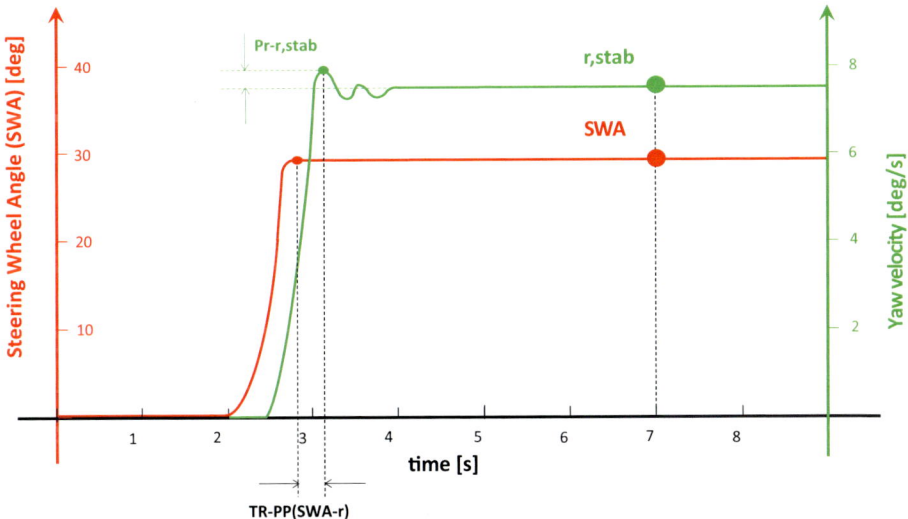

Figure 9 – Trend as function as time of steering wheel angle and yaw rate in a 30 ° step steer maneuver at 100 km/h.

The quantities that are monitored in this maneuver are:
- *TR-PP(SWA-r)*: It is the delay time between the yaw rate peak (we will call it also yaw peak) and the maximum steering wheel angle. The lower the value, the more ready and faster the car is in the change of direction.
- *Pr-r,stab*: It is the difference between the yaw peak and the stabilized value. It depends on the characteristics of the vehicle and the speed at which the steering wheel is operated. The following figure shows the yaw rate trends of three steps steer with three different steering wheel operating times. The green curve relates to a linear ramp of 0.2 seconds, the red one of 0.5 and the blue one of 0.7 s.

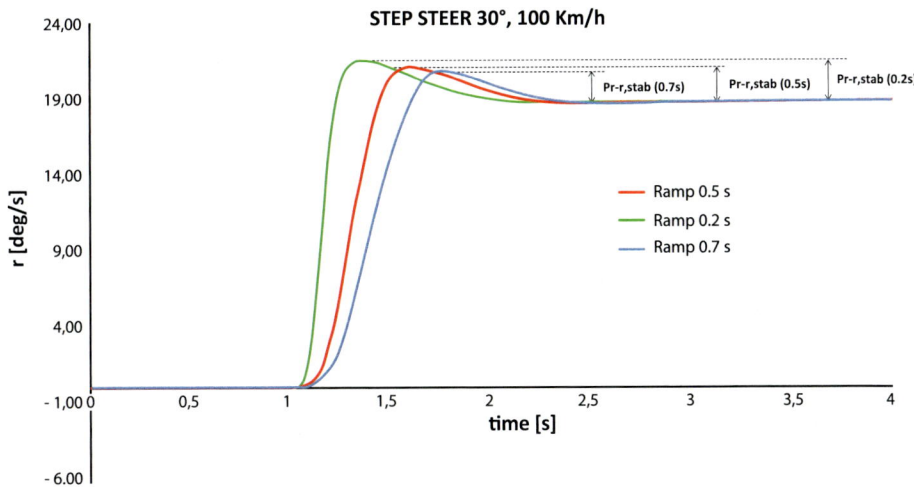

Figure 10 – Steps steer in which the steering wheel speed follows linear trends with duration of 0.2, 0.5 e 0.7 seconds.

209

Automotive suspension

In this case it is evident that the yaw peak increases with the speed of rotation of the steering wheel. In general, however, the yaw peak, or the *Pr-r,stab* value as a function of the steering speed, has an initially increasing trend, reaches a maximum and then decreases. The following figure shows the trend of the yaw peak in a constant speed step steer as a function of the steering wheel operating speed (also called the steering wheel frequency) which in the case in which it is close to 0 (very low) is *r,stab*. As the steering speed increases, the yaw peak increases by reaching a maximum at *Vr,max* and then decreases at higher speeds. By operating, for example, the steering wheel at speed *V**, the yaw rate reaches a peak equal to (*r, stab* + Δ) and then stabilizes at the value *r,stab*, where Δ represents *Pr-r,stab*. This trend is also valid for the other quantities such as the sideslip angle and the lateral acceleration which, after an initial oscillation that generates a peak, stabilize at the stationary value. In sports cars, the max yaw peak, and accordingly the Δ,max, must be low because it represents an undesirable oscillatory response of the rear axle. It manifests itself with a two-stroke yaw response that increases as the steering speed increases and which, in addition to delaying the insertion in cornering, produces a different behavior depending on the speed at which the steering wheel is operated. Therefore, the goal is to flatten the curve in the following figure as much as possible by acting on the slip stiffness of the rear axle. This, if too low, triggers yaw vibrations around the center of gravity due to excessive variations in sideslip angle, as it will be clarified later. The trend of the next curve also depends on the yaw stiffness illustrated in chapter 6.

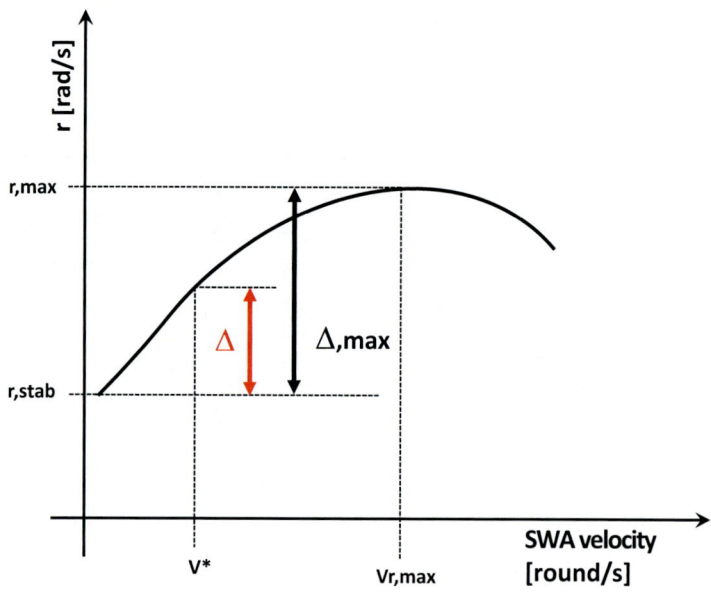

Figure 11 – Trend of the peak of the yaw rate in a step steer of 30 degrees at 100 km/h as a function of the steering wheel operating speed.

- TR-PP(AY-r): This quantity cannot be extrapolated from the previous graph but from the following figure 15. It represents the time difference between the peak of the yaw rate and the peak of lateral acceleration which is always delayed with respect to the yaw. It is measured in hundredths of a second and the lower the sportier the car is. It represents the speed of entry into the curve because it indicates the delay with which the lateral leaning of the car occurs or the delay of the rear axle in finding leaning (achievement of maximum lateral acceleration) in the face of the change of direction. In fact, as will be clarified later, the maximum lateral acceleration comes when approaching the stabilized sideslip angle.
- r,stab/SWA: Also called *Gr* or yaw gain, it is the ratio between the stabilized yaw rate and steering wheel angle. The more this ratio increases, the more the vehicle ability to turn at a certain steering wheel angle increases. The ratio *r,stab/SWA* in linearity conditions (lateral acceleration ≤ 0.4g) is related to the understeer gradient at constant speed *Kus,cv* which is the slope at the origin of curve which represents the steering wheel angle as a function of the lateral acceleration at constant speed, called constant speed understeer curve. The abscissa axis of that curve contains the stabilized accelerations, obtained by carrying out several steps steer at a constant speed by varying the steering wheel angle, while the axis of ordinates contains the corresponding steering wheel angles.

The constant speed understeer curve has a higher slope than the constant radius understeer curve, as seen in the following figure. The constant radius understeer curve is obtained by reducing the kinematic steering angle from the constant speed understeer curve.

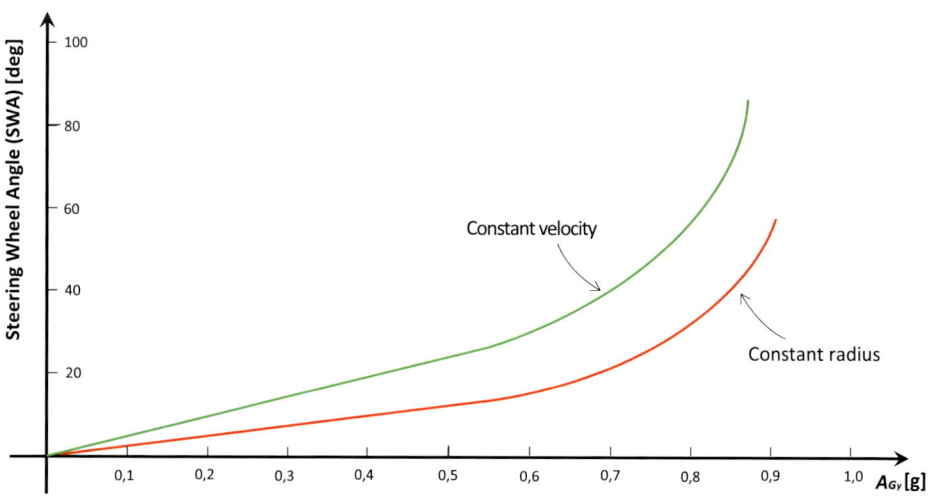

Figure 12 – Constant radius understeer curve (passing through the origin to display the slope comparison) and constant speed understeer curve.

Regardless of the demonstration, it results

$$Kus,vc = Kus + \frac{\tau p}{V_x^2} \frac{180}{\pi}$$

We calculate Kus,cv at 100 km/h

$$Kus,cv(100) \; [deg/(m/s^2)] = \frac{AVOL}{A_{Gy}}$$

and then

$$A_{Gy} = \frac{AVOL}{Kus,cv(100)}$$

Since we are at constant speed and in stabilized conditions (zero sideslip variation), from the formulation of lateral acceleration seen in chapter 4, it results:

$$A_{Gy} = (r,stab)V_x$$

by equating

$$(r,stab)V_x = \frac{AVOL}{Kus,cv(100)}$$

It results

$$\frac{r,stab}{AVOL} = \frac{1}{V_x \, Kus,cv(100)}$$

by inserting Gr equal to the ratio between $r,stab$ and SWA, we obtain:

$$Kus,cv(100) = \frac{1}{V_x \, Gr}$$

Multiplying the second member by 9.81, which is equivalent to replacing A_{Gy} with A_{Gy}/g in the first of these last six equalities, we get the understeer gradient at constant speed in *deg/g*.

All the quantities monitored in the step steer maneuver have target values that depend on the segment to which the vehicle belongs. It is advisable, both in a virtual and an experimental environment, to make parametric comparisons between different vehicle configurations.

8.4.1. Yaw velocity and sideslip angle

As already seen, the yaw rate has a short transient which always manifests itself with a peak and with a slight vibration that subsequently dampens and stabilizes. This peak, which represents the maximum amplitude of the vibration, is the sum of a first contribution that makes the vehicle rotate around the center of curvature and of a second contribution that makes it rotate around the center of gravity, so as to assume the appropriate sideslip angle. The first contribution is very close to the stabilized value *r,stab* while the

Vehicle dynamics – Performance evaluation

second is close to the difference between the peak and the stabilized value (*Pr-r,stab*) or Δ of figure 11 and has an oscillatory trend. The following figure shows a vehicle that passes from configuration A with sideslip angle $\beta A>0$ kinematic (see chapter 4) to configuration B with sideslip angle $\beta B<0$ thanks to the rotation around the center of gravity immediately after the step steer. The center of gravity velocity always remains tangent to the trajectory, the front end moves towards the center of curvature while the rear towards the outside.

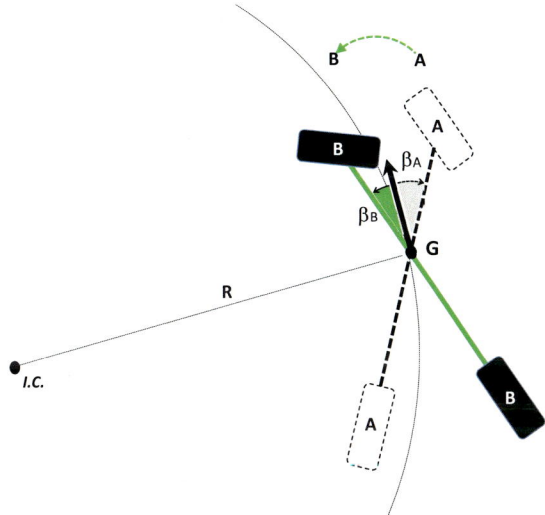

Figure 13 – Rotation of the vehicle around the center of gravity to reach the sideslip angle. A represents the initial kinematic position when the steering wheel angle is imposed and B represents the stabilized position.

The trend of the sideslip angle that arises after the step steer consists of a short transient that stabilizes at a constant value, as shown in the following figure. The interval ΔT necessary to reach the stationary value of the sideslip angle is a very important quantity for the evaluation of vehicle dynamics.

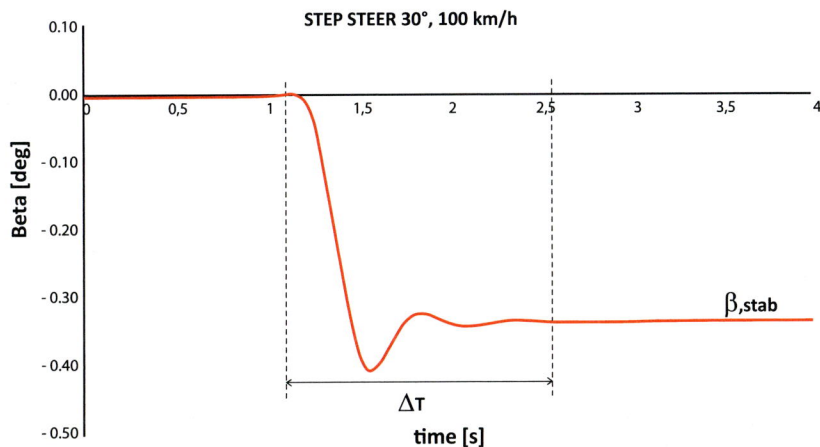

Figure 14 – Trend of the sideslip angle in a step steer at constant speed.

From the observation of the trend of the sideslip angle, and bearing in mind that its derivative over time represents the yaw speed around the center of gravity, we can say, with a good approximation, that:

- As the ratio between the stabilized sideslip angle $\beta,stab$ and the time interval ΔT, necessary for stabilization, increases, the $Pr\text{-}r,stab$, and therefore to the Δ of figure 11, increases.
- The reduction in ΔT is almost always a consequence of the reduction in $\beta,stab$, which stabilizes faster as it is lower.

In conclusion, in order to differentiate the weights of the aforementioned factors (Δ and ΔT), we can affirm that the reduction of $\beta,stab$ moderates the effect of the aforementioned two-time yaw response (because the Δ of figure 11 decreases) while the contraction of ΔT, in addition to reducing the duration of the yaw peak, speeds up the stabilization of the lateral acceleration of the vehicle, giving a performance improvement that will be further clarified in the next paragraph. In light of the fact that the reductions in $\beta,stab$ and ΔT almost always lead to an improvement in performance, the following are the actions necessary to lower them:

1. Increase the slip stiffness of the rear tires;
2. Reduce loss of rear *camber* under lateral load;
3. Increase the rear static *camber* and rear gain in bump travel;
4. Increase the rear static *toe-in* (closed wheel) and the rear *toe-in* under lateral load;
5. Increase the rear *toe-in* in bump travel.

In addition to the first three actions, which, as seen in chapter 2, increase the slip stiffness of the rear axle, the latter also produce the same effect as, with the same lateral force, the slip angle with respect to the vehicle decreases because it is reduced by the closure under lateral load and in bump travel (see chapter 6). It is necessary, however, to make a distinction as the first 4 actions reduce both the $\beta,stab$ and the ΔT while the last one reduces only the $\beta,stab$; it is widely used to lower the constant component in the presence of a bump travel due to weight increase on the rear axle (fully loaded vehicle) in order to make the vehicle more stable.

All the previous actions produce an increase in the frequency of the yaw peak (Vr,max), moving the maximum of the curve in Figure 11 towards the high steering wheel speeds with consequent improvement of the vehicle response (flattening of the curve) at low operating frequencies. The aforesaid actions also lower the value Δ,max of figure 11. The closing of the rear wheel in bump travel and under load represents a good means to increase the peak frequency and to lower the Δ,max. They are used where it is necessary to increase the slip stiffness of the axle without having high-performance tires and suspensions with little loss of *camber* under load. Note that the rear closure in bump travel does not always correspond to a decrease in ΔT, as we will see in the next paragraph. For this reason, on sports cars there is a tendency to increase the slip stiffness of the rear axle through the first four actions and finally through the fifth. This is not possible on cars destined for the transport of people for which the *camber* (both static and dynamic) has the limit of excessive tire consumption, and closing under load is not always possible due to the architecture of the rear suspension.

Rotation around the center of gravity is also inhibited with a lower percentage by:

- Front *toe-out* (wheel opening) under lateral load;
- Front *toe-out* (wheel opening) in bump travel.

which in reality correspond to a lowering of the axle stiffness, because with the same lateral force the slip is higher. The aforementioned front opening increases the value of the peak frequency ($V_{r,max}$) (with less weight than the rear closure) but also slightly increases the value of Δ,max of figure 11. The front opening always produces a delayed response but it becomes necessary when it is not possible to control the slip stiffness of the rear axle, as for example in the case of a twist beam axle that opens under lateral load.

In sports cars it is advisable to lower the front opening as much as possible, both in bump travel and under load, by adjusting the curve in figure 11 only with the rear axle.

8.4.2. Lateral acceleration and sideslip angle

In a step steer at a constant speed, during the rotation of the vehicle around the center of gravity, the lateral acceleration increases and reaches its maximum when the sideslip angle stabilizes. Below is shown the formulation of the lateral acceleration at constant speed. This consists of a first term proportional to the yaw rate and a second term proportional to the variation of the sideslip angle with a duration equal to ΔT of the previous paragraph.

$$A_{Gy} = rV_x + \dot{\beta} V_x$$

The second term, which has a sign opposite to the first, helps to delay the increase in lateral acceleration. It is present only in the first phase in which the car rotates around the center of gravity and is canceled when the sideslip angle stabilizes. When this rotation ends, the acceleration reaches the stationary maximum, and the yaw rate starts towards the stabilized value. In the following figure, the influence of the second member (green curve) with a duration equal to ΔT, during the first step of the step steer, is shown.

Automotive suspension

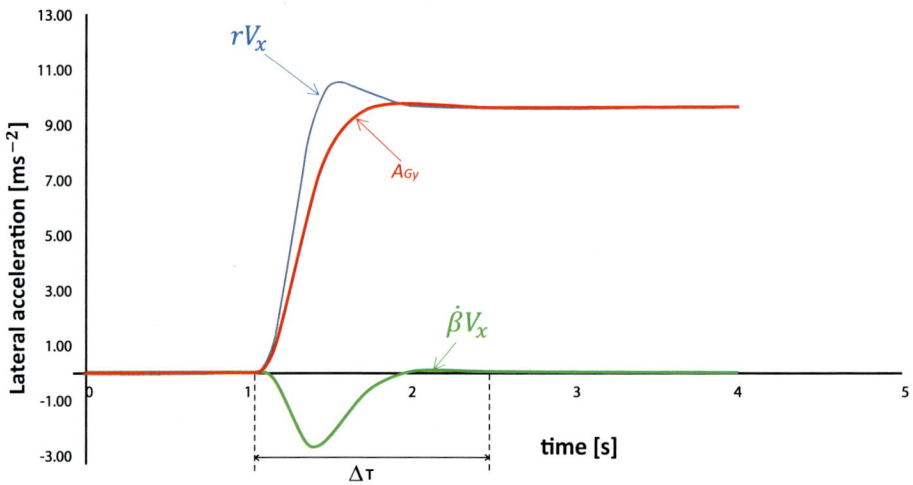

Figure 15 – Trend of lateral acceleration and its two contributions in a constant speed step steer carried out on a racing vehicle.

The ΔT is inversely proportional to the vehicle stabilization speed. In order to quantify the ΔT and its dependencies, we observe that:

- The ΔT decreases as the slip stiffness of the rear axle increases according to points 1,2,3 and 4 of the previous paragraph. This is because the temporal variation of the sideslip angle is inversely proportional to the speed with which the lateral force on the rear axle grows, which is certainly more sudden when the latter is more rigid. Therefore, by increasing the slip stiffness of the rear axle, the sideslip angle and the time interval in question are lowered. If, for example, the rear wheel closure is increased under lateral load, both the stabilized sideslip angle and the range of variability ΔT are lowered and consequently the acceleration peak is anticipated. In the following figure the trend of the sideslip angles of two identical cars with different rear wheel closure under lateral load is shown. The red curve identifies the car that closes the most and has a smaller range of variability as it stabilizes faster

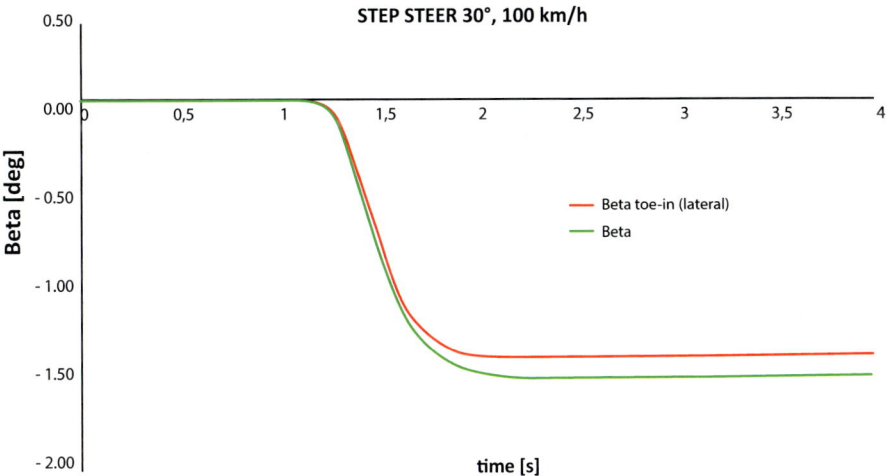

Figure 16 – Sideslip angles, in a constant speed step steer, of two cars with different rear closure under lateral load.

Deriving the sideslip angles we see that the second term of the lateral acceleration formulation dies out faster in the case of the car with greater rear closure. In the following figure, the representation of what is stated with the obvious difference between the two ΔT.

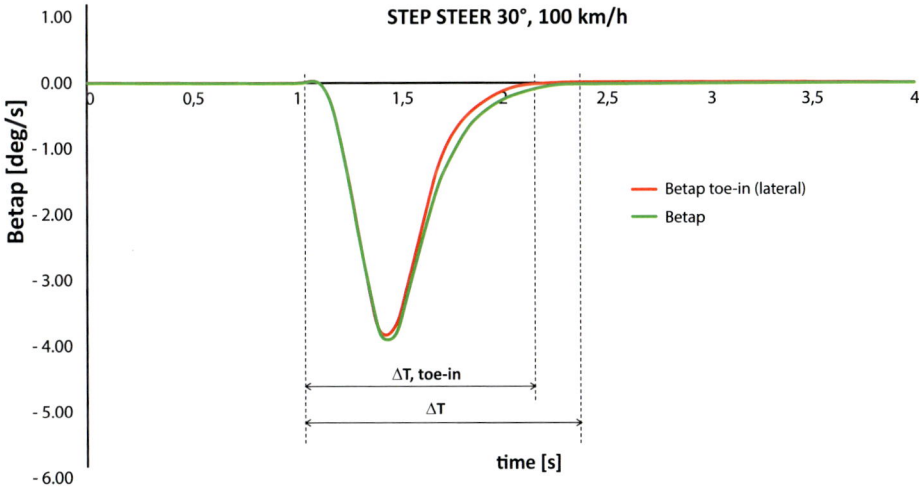

Figure 17 – Derivatives of the sideslip angles, in a constant speed step steer, of two cars with different rear closure under lateral load. The derivative of the car with greater closure dies out faster.

The contraction of the ΔT might seem obvious given that, with the same steering wheel angle, the vehicle with the lower sideslip angle turns with a greater radius of curvature and consequently less lateral acceleration. Instead, in the vehicle with greater rear closure, the sideslip angle continues to be lower even at the same lateral acceleration. In fact, if the step steer at the same velocity is made on the latter with a steering wheel angle that produces the same lateral acceleration, a minor variation in sideslip angle is obtained which contracts the ΔT. In the following figure the comparison between the derivatives of β in which the curve of the car with the transversely stiffer rear axle was obtained by means of a step steer at 31 degrees, in order to equal the lateral acceleration and the radius of curvature in the stationary section.

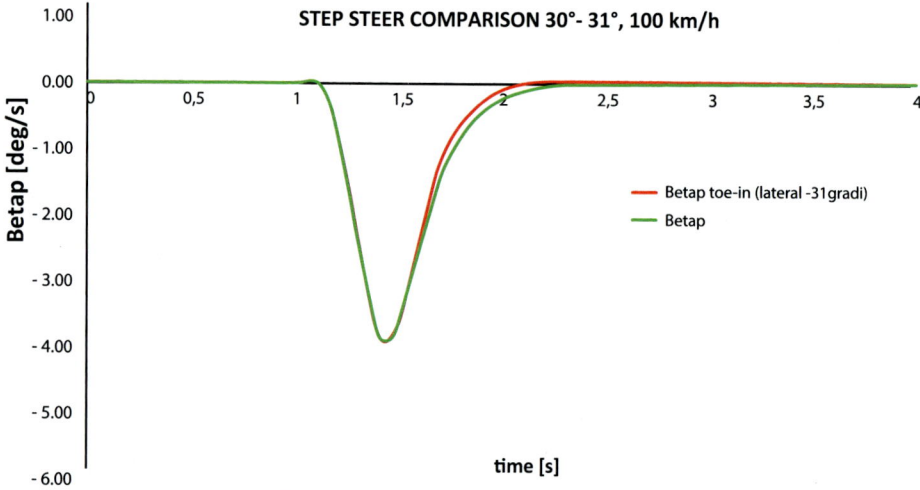

Figure 18 – Time trend of the derivatives of the sideslip angles, in a step steer at constant speed, of two cars with different rear closing under lateral load. In order for the two cars to have the same lateral acceleration, the car with greater rear closure must turn with a steering wheel angle of 31 degrees.

In this case, we can say that, with a set radius of curvature, the car with the lowest $\beta,stab$ is faster to reach maximum lateral acceleration because it has a lower ΔT.

- The ΔT depends on the temporal variability of the sideslip angle and not only on the stabilized value. The decrease in the stabilized sideslip angle does not always correspond to a lower temporal variability of the same and the consequent lowering of the ΔT. This situation can occur in the presence of wheel closure in bump travel which lowers the slip angle of the rear axle at a later time than when it is generated by the lateral force due to acceleration. This happens because the closure of the wheel in bump travel requires that the roll is formalized, which is always delayed with respect to the lateral force. Due to this phenomenon, when the bump travel closure occurs, the modulus of the sideslip angle undergoes a trend reversal, which produces an oscillation that inhibits the contraction of the ΔT. The following figure shows the trends of the sideslip angles of two identical cars which differ only in the rear wheel closure in bump travel. The red curve, which is the one of the cars that closes the most, shows a minimum and a subsequent rise due to the delay just described.

Vehicle dynamics – Performance evaluation

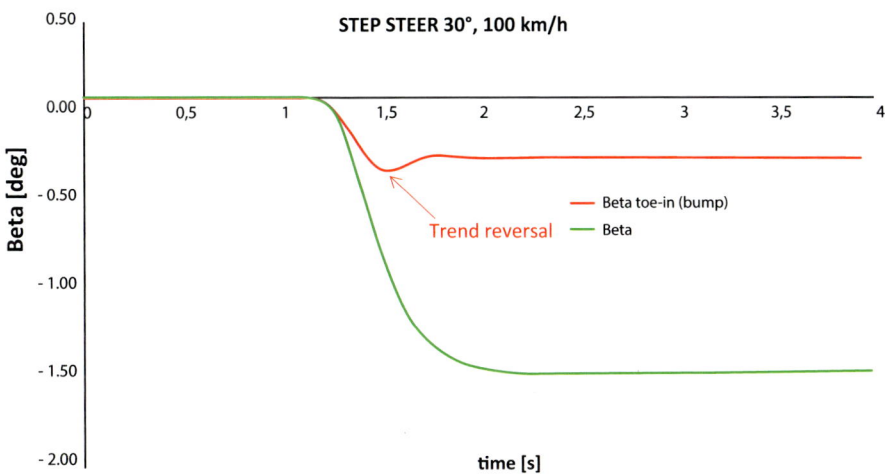

Figure 19 – Temporal trend of the sideslip angles of two car with different *toe-in* in bump travel, in a step steer at constant velocity.

By deriving the sideslip angles, the trends shown below are obtained. In these the oscillatory response of the red curve is evident, which does not allow the contraction of ΔT. It is equally evident that the greater rear wheel closure in bump travel reduces the yaw peak as it lowers the contribution of the second member of the lateral acceleration.

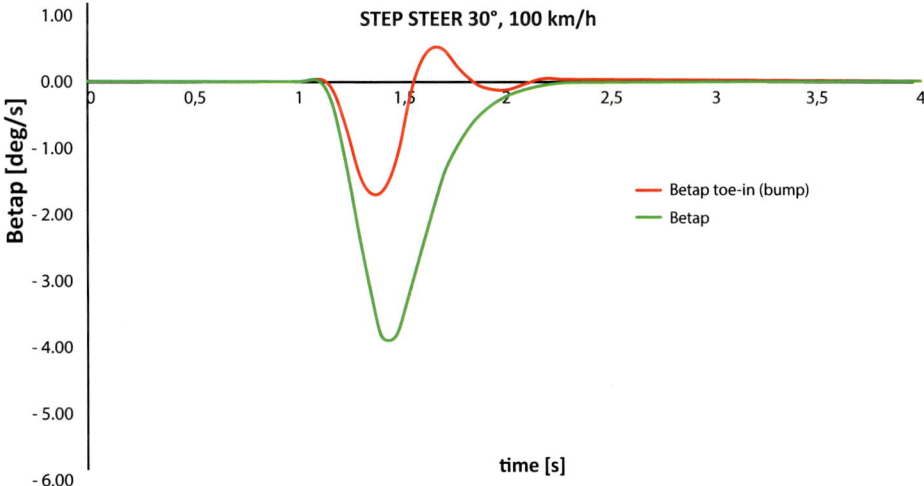

Figure 20 – Derivatives of the sideslip angles, in a constant speed step steer, of two cars with different rear closing in bump travel. The derivative of the car with greater closure has an oscillatory behavior.

The situation just described occurs when the wheel closure in bump travel produces an important reduction of the axle slip.

- The ΔT decreases when the speed with which the lateral force on the rear axle increases. In particular, the *toe-in* under lateral load and the *camber* gain in bump travel on the rear axle produce an increasing surplus of force directed towards the inside of the vehicle which speeds up the achievement of the maximum required lateral thrust and its stabilization. Since in the transitory the lateral force request on the rear is proportional to the term $J\dot{r}$ of (4.15), it is clear that a more contained moment of inertia J lowers the peak of force required and allows the contraction of the $\varDelta T$. The *camber* and the static *toe-in* also produce a preload towards the inside of the vehicle which helps to reach earlier the peak of force required at the rear and to contract the $\varDelta T$.

- As $\varDelta T$ increases, the distance between the acceleration peak and the yaw peak-increases, and then the delay TR-PP(AY-r), proportional to the vehicle stabilization speed. This is because the yaw peak always corresponds to the peak of the first term of the formula which is always around the center of the $\varDelta T$, while the acceleration peak corresponds to the zeroing of the second term which is always around the end of the $\varDelta T$.

IT is important to reiterate that the vehicle stabilization speed is not the vehicle speed during the curve maneuver, but the speed with which the dynamic quantities stabilize that leads the vehicle to turn according to the minimum radius of curvature. A vehicle with the $\varDelta T$ contracted is certainly easier to maneuver because it stabilizes more quickly but, for the same steering wheel angle, it could turn less quickly than a second vehicle with the expanded $\varDelta T$. This happens when the expansion of $\varDelta T$ depends on a greater sideslip angle (lower slip stiffness of the rear axle) which lowers the final radius of curvature and increases the yaw rate. Maximum performance is obtained by increasing the steering ratio on the first vehicle in order to obtain, with the same steering wheel angle, the increase in yaw speed and the decrease of the radius of curvature with a higher level of maneuverability. A vehicle with $\varDelta T$ contracted arrives more suddenly at the maximum lateral acceleration and consequently at the minimum radius of curvature, defined by the following approximate formula, valid only after the cancellation of sideslip angle variation. The more sudden reaching of the minimum radius anticipates the transverse displacement and brings the vehicle at $\varDelta T$ contracted to turn earlier.

$$R = \frac{V_x^2}{A_{Gy}}$$

Similarly, with a set trajectory, the achievement of the target radius of curvature (imagined by the user when setting the steering wheel angle) is faster for the vehicle with the contracted $\varDelta T$.

For these reasons, the important sideslip gain of some cars, such as those with the twist beam axle, is an undesirable behavior both for the stabilization speed and because the excessive slip of the rear axle brings the vehicle closer to instability, making the car difficult to manage in extreme situations. However, the twist beam axle can be a good way to make the car more direct in slalom or in curves with low radii of curvature, if a high steering ratio is not available (steering gear not very direct). In fact, the greater slip of the rear axle, with the same steering wheel angle, lowers the radius of curvature and the pronounced rotation around the center of gravity shifts the direction of the car towards the center of the curve, giving the driver the feeling of having well entered with the vehicle in between the cones. In addition, the increase in yaw rate, resulting from the decrease in the radius of curvature induced by the sideslip angle, could be an advantage in a condition, such as the slalom, where the vehicle does not have time to stabilize and therefore to perceive a low *TR-PP(AY-r)*.

In conclusion, we can say that in very fast cars, in order not to make the transient cornering too "nervous", the sideslip angle must be low and with the smallest possible variability. However, the sideslip angle must always be present, otherwise the slope of the understeer curve would increase too much in the last section, with a consequent increase of *S* area and lowering of maximum acceleration. Furthermore, it is important to remember that the excessive difference between the yaw peak and the stabilized value *(Pr-r,stab)* accompanied by the vibration, which occurs as the sideslip angle and its variability increase, indicates a different reaction of the vehicle depending on the speed at which the steering wheel angle is imposed. Generally, this is an undesirable behavior as it confuses the user who perceives a difference in behavior depending on how he operates the control.

Below is a summary of the factors that influence the yaw response trend of figure 11 and the ΔT:

- *Vr,max*: Increases through actions 1,2,3,4,5 (previous paragraph) and through the front wheel *toe-out* (in bump travel and under lateral load) with less weight than rear *toe-in* of actions 4,5.
- *Δ,max:* It is lowered by actions 1,2,3,4,5. It increases slightly with the front wheel *toe-out* (in bump travel and under lateral load).
- *ΔT*: It is lowered through actions 1,2,3,4 because they lower the sideslip angle and its variability.

8.4.3. The relaxation length

Another important dependence of the yaw peak trend, as a function of the speed of actuation of the steering wheel, lies in the relaxation length of the tires, on which we will give a very qualitative explanation. This is the length in meters that the wheel must travel in order for the force to reach approximately 63% of its stabilized value when the slip is imposed. If the relaxation length increases, the response delay of the lateral force of the associated axle increases and consequently the response of the car changes. If, for example, we increase the front relaxation length, when the rear axle acquires maximum force, the rotation around the center of gravity is inhibited by the delay of the force on the front axle which slows the gain of sideslip angle. In this way, the greater moment of the rear force will cause the car to reach a lower sideslip angle peak and consequently a lower yaw rate peak. When the two forces will reach the maximum value, the yaw rate and the sideslip angle will assume the steady state values, different from the peak values assumed during the transient and not depending on the relaxation length.

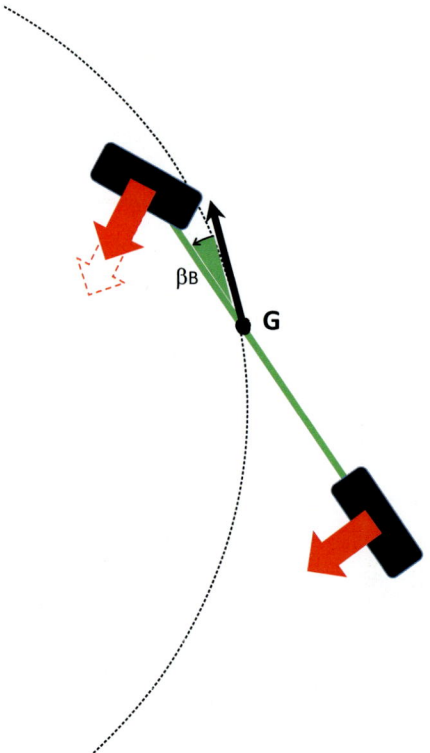

Figure 21 – Delay of the front force compared to the rear due to the greater relaxation length on the front axle.

A similar effect is obtained by decreasing the relaxation length on the rear axle. These are two other ways to lower the Δ,max of figure 11.
Many sports cars have different tire sizes on the two axles and therefore also have different relaxation lengths. Generally, the relaxation length decreases as the ratio between width and height of the tire shoulder increases. Furthermore, in most cases, the relaxation length decreases as the pressure increases and also depends on the ratio between the frequency imposed on the tire and its forward speed. With the same pressure, as this ratio increases, the relaxation length increases, and thus as the speed increases, the relaxation length decreases. However, during the tuning phase it is possible to try rising the pressure at the rear and lower it at the front in order to flatten the curve in figure 11. Obviously, it is advisable to apply small variations within the allowed range in order not to also vary the lateral stiffness.

8.4.4. The roll angle

Another vibratory behavior to keep under control is the one of the roll angle, which is an excellent indicator of the correct sizing of the shock absorbers. The car in curve 1 in the following figure, after the step steer undergoes an oscillation before stabilizing at

Vehicle dynamics – Performance evaluation

about 4 degrees. This is an undesirable behavior. In addition to being perceived in the passenger compartment by the user as the vehicle body oscillates annoyingly before stabilizing, it is also perceived on the steering wheel. This is because the vibration of the roll corresponds to the travel of the suspensions, with the consequent variation of *camber* and *toe* that go to influence the direction of the vehicle. The effect of the aforementioned vibration is also manifested by the increase in the variability of the yaw rate and the sideslip angle that produce the expansion of the $\varDelta T$.

Car 2 progressively reaches the limit value because its shock absorbers are correctly sized as illustrated in chapter 3.

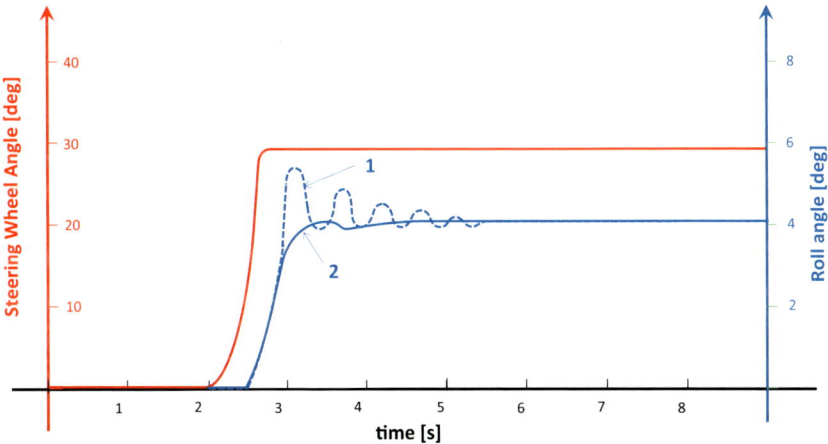

Figure 22 – Roll angle trend with two different shock absorber set-ups in a 30 ° step steer maneuver at 100 km/h.

8.4.5. Considerations on different behaviors

In light of all the observations made, we can give an interpretation to the three different behaviors in figure 8 which we propose again below for convenience.

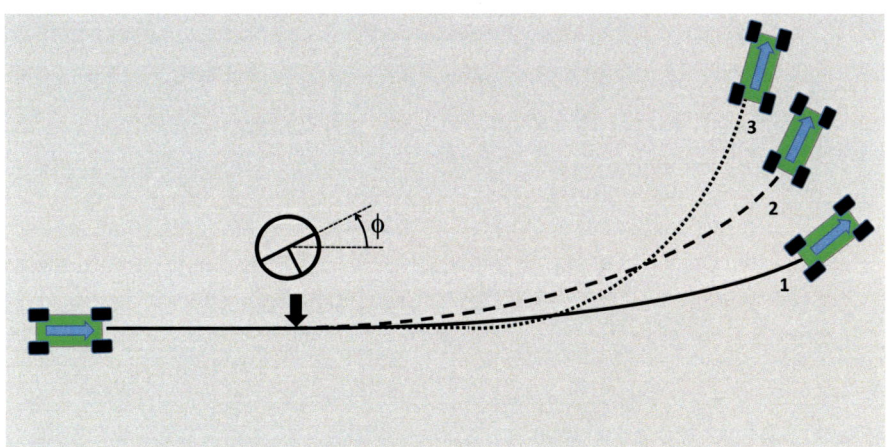

Figure 23 – Three different behaviors in a step steer maneuver.

Automotive suspension

Below are the reasons for the different trajectories.

1. Car 2 may have a higher steering ratio than 1. Therefore, car 2, with the same steering wheel angle, has a lower radius of curvature.

2. Cars 1 and 2 may have different suspension schemes and therefore different *Kus*. Car 2, with the same steering wheel angle, produces greater lateral acceleration and therefore a smaller radius of curvature at constant speed.

3. The TR-PP(SWA-r), namely the delay between the yaw peak and the maximum steering wheel angle, is smaller for car 2 (compared to car 1) which, at the moment in which the image was frozen, is more transversely placed because it started to turn earlier. In this case, the lowering of the delay depends a lot on the slip stiffness of the front axle, on the set-up of the suspensions (bars, springs and shock absorbers) and on the elastokinematics. Surely a car which diverges (*toe-out*) on the front axle (under lateral load and in bump travel), even if more stable, has a greater delay than one that does not diverge. The same happens for a car that has low static *camber* on the front and little gain (bump travel and steering) compared to one that has more. In general, a low value of the *TR-PP (SWA-r)* indicator is a consequence of a low value of *TR-PP(AY-r)*.

4. Car 3 may have a *Pr-r,stab* larger than car 2, because of the greater sideslip angle due to a different rear suspension set-up. As shown in the previous paragraphs, the higher value of sideslip angle (caused by the lower slip stiffness of the rear axle) lowers the radius of curvature. The decrease in the radius of curvature, although it speeds up the maneuver, certainly makes the car more difficult to manage. This is due to the excessive vibration around the center of gravity which can manifest itself to the user with a slight variability of the effort to the steering wheel. Car 3, with the same steering wheel angle set, is faster than car 2 only if the user manages to dominate the effects of this variability which is sometimes difficult, so that the set-up of car 2 is preferred. The greater transverse displacement initially acquired by car 2, graphically emphasized in order to highlight the phenomenon, is the effect of the lesser delay in stabilization which anticipates reaching the minimum radius of curvature. This highlights the greater readiness and progressiveness of response of the car 2 compared to the 3, which is clearly more nervous because it starts cornering late and more abruptly.

5. If cars 1 and 2 are identical, the difference in behavior depends on the speed at which the steering wheel is actuated from which the Δ of figure 11 arises and the consequent peak of sideslip angle which in turn anticipates the curvature of vehicle 2. In this case, since the stabilized values of the yaw rates are the same, the cars acquire the same final radius of curvature.

8.5. Summary

The maneuvers described for the dynamics vehicle dynamics evaluation are:

- *Steering pad*

  Suitable to evaluate the vehicle behavior in stationary. In this maneuver the understeer and sideslip curves are built which are indicators of the operating range of the tires of the front axle and rear axle. In particular, the sideslip variation, which depends on the rear axle, is sized based on the desired shape of the non-linear section of the understeer curve. The slope and length of the linear section of the understeer curve depend on the static values of the characteristic angle and the elastokinematics of suspension. For example, greater front wheel opening increases both the *Kus* and the *AYLIN* while greater front *camber* gain lowers the *Kus*.

- *Step steer*

  Suitable for determining the transitory behavior and evaluating the trend of the response delays and of the waveform of yaw rate which represent the indicators of the speed at which the vehicle turns. The yaw rate trend as a function of the speed of application of the steering wheel angle, represents an indicator of the dynamic performances. The flattening of the aforementioned curve, which gives to the vehicle a sportier response, is carried out by intervening on the slip stiffness of the axles and on the relaxation lengths of the tires. The increase in speed stabilization of the sideslip angle speeds up the achievement of the minimum radius of curvature and anticipates the transversal displacement with the same applied steering wheel angle. Through the step steer it is possible to verify the correctness of the damping of the shock absorbers thanks to the analysis of the roll angle trend.

9. VEHICLE DYNAMICS
Achievement of the targets

We shall now analyze the maneuvers proposed in the first chapter through the background acquired from the understanding of the text, and we will refer to the corresponding chapters for a more in-depth explanation of physical phenomena. In this way the applicative aspect is extracted from all those apparently theoretical topics that have been previously treated. The maneuvers are the following:
- Straight line driving;
- Cornering;
- Slalom.

9.1. Straight line driving

The following are the different straight line driving conditions.

Straight line driving at constant speed

To ensure constant speed, longitudinal acceleration must be zero. This condition is guaranteed only if the longitudinal forces of propulsion, braking and aerodynamics are balanced. The two problems that can arise in straight line driving, regardless of the sign of longitudinal acceleration, are the misalignment of the steering wheel and the lateral vehicle drift. We distinguish the two phenomena below.

Steering wheel misalignment is a kinematic condition that manifests itself with the need to apply a steering wheel angle other than zero to travel straight ahead (the test is carried out at a constant speed to avoid other influences). This phenomenon depends on the asymmetry of the static *toe* on the two axles, and is characterized by the fact that, after the correction, the vehicle is in dynamic equilibrium and the steering wheel angle does not return any reaction torque. As seen in chapter 6, the convergence on the front axle affects more than that on the rear axle which tends to self-align in some cars equipped with elastic bushings and positive *caster*.

The lateral vehicle drift is a directional instability caused by the generation of forces that rise up the steering kinematics chain generating a stress on the steering wheel that forces the user to the undesired application of a steering wheel torque to keep the vehicle in trajectory. This phenomenon depends on:

- *Caster*

    If the *caster* is not symmetrical, the stabilizing actions seen in chapter 7 are not specular and consequently the loads which rise up the steering kinematics are asymmetrical. In particular, the static *camber* and the static *toe* generate transverse forces. These, in the presence of non-symmetrical *caster trail* (caused by the asymmetry of the *Caster*), produce reactions which, reported to the steering wheel, are formalized with a non-zero torque that the user must counter to keep the vehicle in a straight direction. In the following figure, the schematization of the phenomenon with the detail of the torques *Ml* and *Mr* deriving from the two symmetrical lateral thrusts (*Flat*) due to the negative static *camber* and the negative static *toe* (open wheels) is shown.

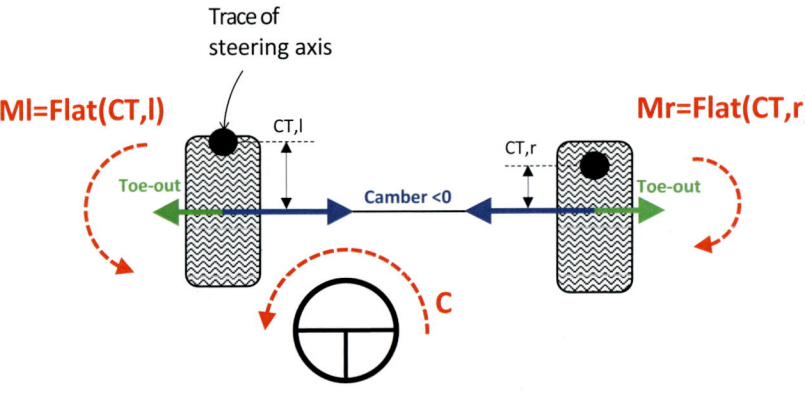

Figure 1 – Example of asymmetrical caster trail. The lateral force (Flat) is the sum of the thrust due to the negative static *camber* and the thrust due to the static *toe-out*. The torque at the steering wheel C, which the user must contrast to keep the vehicle on the trajectory (nothing in symmetrical conditions), is the consequence of the difference between Ml and Mr due to the difference in the caster trail.

When, on the other hand, the *caster trail* is symmetrical but the *caster* is asymmetrical, the *second stabilizing action* seen in chapter 7 is asymmetrical and consequently the reaction torques to the aforementioned transverse forces are different. The reaction torque acting on the lower caster wheel stresses the steering kinematics more because it is not absorbed by the vertical elasticity of the tire, given that, compared to the other wheel, it has less tendency to interpenetration towards the road level.

- *King Pin Inclination*

    The asymmetry of *KPI* is responsible for the asymmetry of the scrub radii. These, multiplied by the resistance to advancement contribute to the generation of two torques *Ml* and *Mr*. In conditions of excessive asymmetry, they do not cancel each other out as they should, but rise up the steering kinematics, manifesting themselves with an unwanted torque at the steering wheel. The following figure shows an example of an asymmetrical scrub radius with the representation of the torque on the steering wheel C that is generated due to the different moments on the wheels due to the resistance to advancement. The phenomenon is similar to that described in chapter 7 concerning braking.

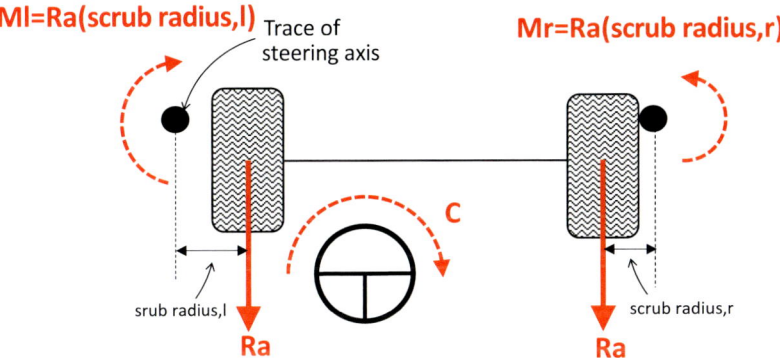

Figure 2 – Example of negative and asymmetrical scrub radius. The torque at wheel C, which in symmetry conditions would be zero, is the consequence of the difference between Ml and Mr due to the difference in the scrub radii.

- *Camber*

   Excessive asymmetry produces a different transverse thrust which tends to move the vehicle towards the wheel with a lower absolute value of *camber*, generating the consequent undesirable reaction at the steering wheel which, if not counteracted by the realignment induced by the caster trail, induces lateral displacement. The asymmetry of the caster can emphasize this effect.

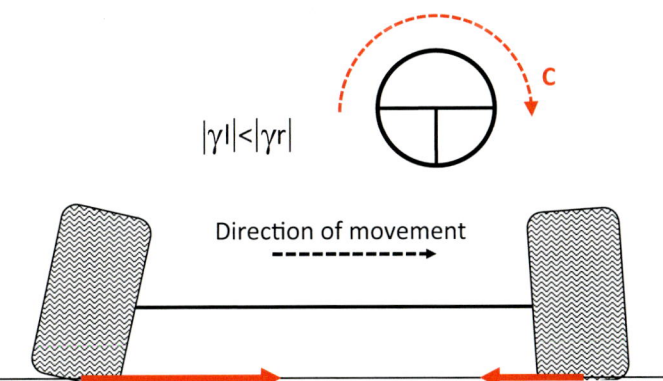

Figure 3 – Front view of a vehicle with the asymmetrical *camber* on the front axle with evidence of the consequent direction of movement of the vehicle and the consequent torque at the steering wheel.

- *Tire asymmetry*

   Different compounds between right and left could generate asymmetrical resistance to the advancement and consequent torques on the wheels that do not cancel each other out.

In the misalignment of the steering wheel, unlike the lateral drift, once a steering wheel angle other than zero has been set, the vehicle continues straight travel without any force that tends to make it change direction. On the other hand, in the drift phenomenon, the force that tends to move the vehicle transversely is always present.

Straight line driving at increasing speed
To ensure increasing speed it is necessary that the longitudinal acceleration is aligned with the direction of travel and this happens thanks to the drive torque. The problem that can arise in the presence of acceleration, and as we will see also under braking, is the drift of the vehicle.

In this case, in addition to the caster, the *camber* and the static *toe*, which influence the maneuver for the same reasons as the previous one, there are other factors that determine the vehicle behavior, which are:

- *Toe* variation under longitudinal load
  For both front and rear traction, the *toe* variation must be symmetrical so that the car does not suffer drift in acceleration. Otherwise, the vehicle would tend to steer according to the orientation taken by the wheels. The ideal situation is the wheel opening if the driving axle is front and closing if the driving axle is rear. But unfortunately, these objectives, which give the vehicle stability even when cornering, are not always achievable due to the compromise with other performances, such as the change in *toe* under braking which must have the same behavior with an opposite load. For a better understanding, refer to the paragraph relating to the change in *toe* under longitudinal load in Chapter 6.

- *KPI* and *KPIoffset*
  As shown in chapter 6, the longitudinal force at the center of the wheel corresponding to the traction, produces two opposing torques on the wheels which, if equal, do not generate any effects on the steering. Otherwise, if the two torques are different due to the asymmetry of the *KPI* and the *KPIoffset*, the difference rises up the steering kinematics and manifests itself with reaction torque at the steering wheel that tends to move the vehicle transversely. Below is the same figure shown in chapter 6 which highlights the moment M which is approximately equal to the product between the longitudinal force F and the *KPIoffset*.

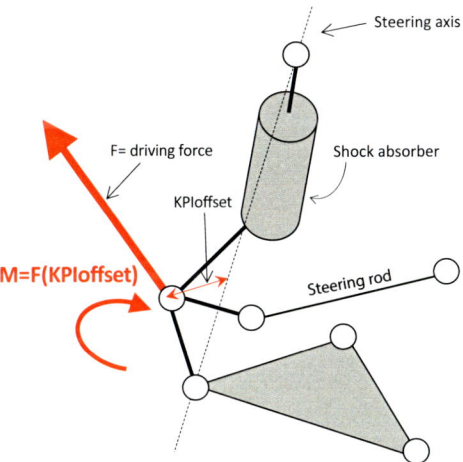

Figure 4 – Left front suspension loaded with a longitudinal force at the center of the wheel that simulates traction.

Straight line driving at decreasing speed

To ensure the decreasing speed it is necessary that the longitudinal acceleration is discordant with the direction of travel, and this happens thanks to the braking torque. In this case the quantities to be kept under control, which affect the drift for the same reason as shown in the traction maneuver, are:

- *Toe* variation in braking.

  In this case as well, in order to avoid vehicle steering, it is necessary that the *toe* variation under braking is as symmetrical as possible. A difference in braking divergence between the front wheels, equal to 6 primes reported to the steering wheel (dividing it by two and multiplying it by the steering ratio, which we imagine equal to 15) corresponds to 0.75 degrees. Such a small difference is equivalent to applying a steering wheel angle of 0.75 degrees at the speed at which you are braking which, if it is on average high, corresponds to making a lane change. The ideal situation is the *toe-out* (divergence) on the front axle and the *toe-in* (convergence) on the rear axle. But as for the variation in traction this is not always achievable. Chapter 6 describes the physical characteristics that the front suspension must have in order for it to diverge under braking. An asymmetrical *toe* variation under braking (both in bump travel and under longitudinal load), in addition to generating the steering of the vehicle, also causes asymmetrical lateral thrusts which, as already seen previously, produce an undesirable reaction torque at the steering wheel.

- *Camber* variation in braking.

  The asymmetry of the front *camber* variation during braking, mainly influenced by the bump travel of the suspension, determines asymmetry of the lateral thrusts that produce, as for the *toe*, an undesirable reaction torque at the steering wheel. In general, the presence of asymmetrical lateral thrusts produces an opposite asymmetry on the maximum braking forces. Based on the conformation of the adherence ellipse, this phenomenon manifests itself with a greater longitudinal force on the wheel with less lateral thrust. The difference between the maximum braking forces produces, in extreme conditions, an additional yaw torque, which has the same direction as that induced by the difference in *camber*. Furthermore, in the case of sliding, the lower longitudinal force is further lowered (transition from static to dynamic friction), emphasizing the yawing effect. In the following figure the graphic representation of the yaw torques due to the transverse thrusts and braking forces is shown.

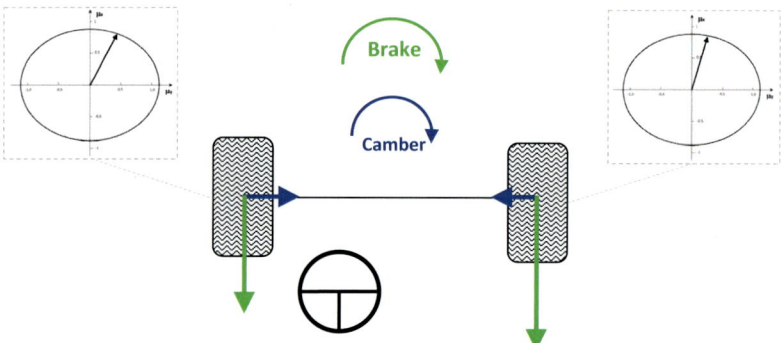

Figure 5 – Plan view of the front axle with the transverse thrusts, due to the *camber*, and the maximum longitudinal braking forces in adherence conditions. The components of the forces on each tire derive from the working point on its adherence ellipse.

- *Scrub radius*

  The scrub radii, if not perfectly symmetrical, are responsible for the asymmetry of the torque at the wheels which, when rise up the steering wheel via the steering mechanism, produce an annoying rotation that the user must counteract to stay on the trajectory. The description of the phenomenon is shown in the figure below. In chapter 7 the influence of the scrub radius in braking on asymmetrical road surfaces has been extensively described.

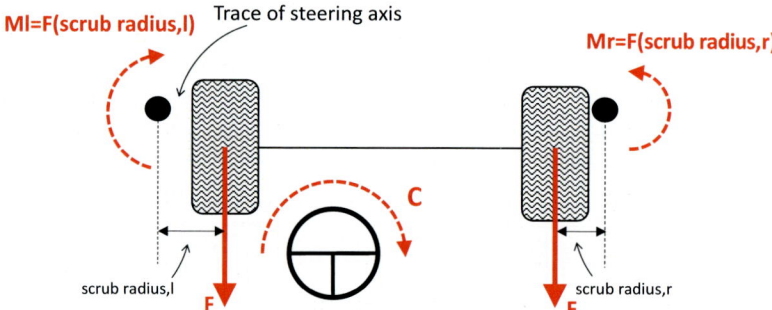

Figure 6 – Plan view of the front axle with asymmetry of the scrubs radii which, in braking maneuver, cause the torque at the steering wheel C due to the difference Ml-Mr.

· Symmetrical behavior of the braking system.

  It is quite intuitive that if the braking capacity of the left side is different from that of the right side, the corresponding ground forces are asymmetrical and produce

a yaw torque with the consequent lateral drift of the vehicle.
- *Caster* and *KPI* variations in parallel wheel travel.

  The asymmetrical variation of these quantities, resulting from the extension (acceleration) and compression of the springs (braking), produces asymmetrical response. If the static values are symmetrical and the car complains of lateral drift, it is advisable to check that the variations in parallel wheel travel are also symmetrical.

All the above causes cannot be quantified precisely because they must be contextualized in a specific vehicle. The general rule is that, for the car to maintain directional stability, the suspension must be as symmetrical as possible. Where this condition cannot be achieved due to industrial processes and production volumes, the following tolerances (with a steering ratio of around 15) can be considered acceptable:
- +/- 4': static *toe* on single wheel;
- +/- 4': difference between *toe* left and *toe* right under maximum braking load;
- +/- 30': static camber, caster and *KPI*.

The tolerances of the static convergence are sized in such a way as to produce a steering wheel misalignment that is not perceived by the average user.

In each of the three maneuvers, the user expects the vehicle to maintain the direction of travel set with the steering wheel fixed or straight, even in the presence of random external disturbances that cannot be controlled by the driver. For this reason, as seen in chapter 6, it is important that the vehicle has a good yaw stiffness that allows to automatically react to external perturbations.

9.2. Cornering

We divide the curve into the four phases shown in the following figure

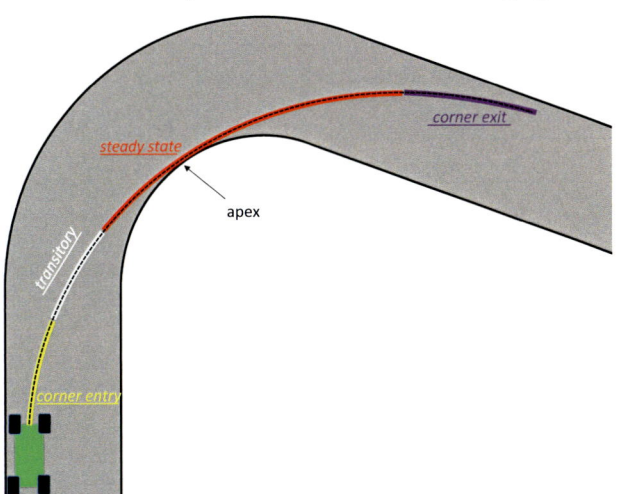

Figure 7 – Fictitious division of the cornering maneuver into four phases which are normally overlapped. The figure highlights the apex which is generally located near the center of the curve. The size of the phases is not real but is only indicative.

Corner entry

The *corner entry* phase consists in rotating the steering wheel by the amount ϕ that the user deems appropriate to face the curve or to direct the vehicle towards the apex.

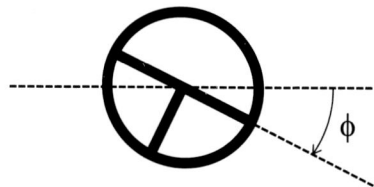

Figure 8 – Steering wheel angle set in the first phase of the curve.

At the end of the application of the steering wheel angle, the lateral acceleration is still low and the load transfers have not yet been formalized but the vehicle has already undergone a transverse displacement. The next phase, called transitory, starts from the moment in which the suspensions begin to change configuration thanks to the presence of the load transfers. It should be remembered that the subdivision was made for clarification purposes only and that there is not a discrete transition between the two overlapping phases.

The value of the steering angle of the *corner entry* phase derives from the slip stiffness of the axles in static conditions and from the steering capacity of the vehicle. We shall now describe where to act to change it.

- Steering ratio.
  The steering ratio is the factor that most influences the steering wheel angle in the *corner entry* phase. For this reason, one of the first features supplied to sports cars is a low steering ratio.

- *Ackermann* percentage.
  The *pro-Ackermann* steering, in addition to lowering the steering wheel angle because it acts on the steering ratio, helps to make the car more ready and faster in the change of direction because it produces favorable dynamic reactions when cornering. As discussed in chapter 7, the *Ackermann* percentage can increase in front-engined cars, in the event of large aerodynamic deportations and when the inner tire does not undergo significant lowering of the *sliding limit* thanks to the elastokinematics of the suspension.

- Static *camber* on two axles.
  As seen in chapter 2 and later in chapter 6, the static *camber* increases the slip stiffness of the axles and consequently affects the K_{us} and therefore the steering wheel angle. In particular, the steering wheel angle in *corner entry* decreases both if the front static *camber* is increased and if the rear one is decreased. Generally, racing

cars have higher *camber* values on the front axle than the rear to make the car not only more direct when cornering but also more reactive. Road cars, on the other hand, have a greater rear static *camber* than the front one to have greater understeer. As seen in chapter 6, to adjust the *Kus* and the initial steering wheel angle, it is important to calibrate the difference between front static *camber* and rear static *camber*.

The dynamic influence consists in the fact that, by increasing the front static *camber*, since the characteristic of the tires translates to the left (see chapter 2), with the same slip angle imposed on the wheels (and therefore the steering wheel angle ϕ), the lateral force on the front axle which determines the yaw of the vehicle increases. The static *camber* on the front axle can also be increased through the *camber* gain in steering due to the *caster*. In this way, as the *caster* increases, the *camber* with which the car faces the first part of the curve increases.

• Static *toe* on two axles.

As explained in chapter 6, since front static *toe-out* (diverging wheels) makes the vehicle faster in sudden changes of direction, it is advisable to adopt a set-up of this type to make the steering more reactive. Keep in mind that this is a dynamic effect, due to some forces that tend to make the vehicle yaw in accordance with the curve, which occurs only in the *corner entry*. On the other hand, when the roll travel is completed and the vehicle has stabilized, the required steering angle becomes greater than in the case of static *toe-in* (converging wheels). The rear axle contributes to lowering the steering wheel angle with *toe-out* (diverging wheels) but this is a set-up that is generally adopted only on slalom sports cars. This is because, although it guarantees great agility in tight corners, it causes instability of the vehicle especially at high speed as it contributes to lowering the yaw stiffness described in chapter 6.

• Slip stiffness of the tires.

The increase in the slip stiffness of the tires lowers the slip angles. However, this does not always correspond to the decrease in the steering angle because it depends on the difference between the axles, as shown in chapter 4. To lower the initial steering angle, the slip stiffness of rear tires must be lowered. Alternatively, the stiffness of the front tires must be increased. To lower the rear, the width can be reduced or the height increased with the same radius under load, while the front can be increased by increasing the width or reducing the height. This might seem contradictory with regard to the common orientation of sports vehicles which tends to adopt greater widths on the rear. This last trend serves to guarantee understeer behavior (in stabilized) to vehicles with mass distribution shifted towards the rear or to very powerful rear-wheel drive vehicles, that want to limit the sideslip angle in cornering acceleration both to have less lateral acceleration delay and to avoid risk oversteer in extreme maneuvers.

- Longitudinal arrangement of the masses.
  The angle ϕ increases as the percentage of weight on the rear axle increases.
  The vehicle ability to enter a curve depends on the value of the front lateral force that it is able to exert with a given steering wheel angle. The application of the steering wheel angle produces two slip angles which, depending on the characteristics of the front tires, give rise to a final value of lateral force which increases as the vertical load on the axle increases. For this reason, forward-shifted weight distribution requires less tire slip angle, and consequently less steering angle, to obtain the transverse force capable of causing the car to yaw. This trend in *corner entry* contrasts with the stabilized one according to which the steering wheel angle decreases when the weight distribution moves towards the rear axle. The latter behavior is formalized with a minor $\Delta\phi$ which lowers the total steering wheel angle (in case of percentage shifted to the rear).
  The *caster* increase is a way to make vehicles with little vertical load on the front axle more responsive in *corner entry*.

- Moment of inertia and wheelbase.
  The angle ϕ increases as the moment of inertia around the z increases.
  In order to achieve a certain angular acceleration, the maximum value of the lateral force required to the front axle decreases as the moment of inertia *J* around z decreases, as explained in the last paragraph of chapter 4. The expression below of the force in which the inverse dependence on the vehicle wheelbase is also noted:

$$Fyi, f\,[N] = \frac{J\dot{r}}{p}$$

Consequently, with a more contained *J*, the angle ϕ in *corner entry* decreases because the achievement of the desired angular acceleration requires less lateral force. The reduction of *J* is carried out by bringing all the masses closer to the longitudinal line and the transversal line passing through the center of gravity of the vehicle. The shift of the driver's seat towards the center, the adoption of suspension schemes with push rod and the use of lighter rims are examples of *J* reduction. The increase in wheelbase also makes the car more direct when entering, since it lowers the lateral force request, but must not lead to an increase in the moment of inertia.

All these features contribute to defining the steering wheel angle related to the *corner entry*.

Transitory

It represents the time interval in which the vehicle reaction evolves following the application of the initial steering wheel angle ϕ and ends when the correction $\Delta\phi$ is completed. In this phase, the lateral acceleration reaches its maximum and the vehicle rolls, making the suspensions acquire a new configuration. While the wheels take on successive new orientations, due to the bump-rebound travel and the elastic deformations of the bushings, the user corrects the steering wheel angle of the quantity $\Delta\phi$ to stay on the trajectory; this correction is not instantaneous but progresses as the roll increases. In order to evaluate the trend of the dynamic quantities and to estimate the time to conclude the transitory, we define the temporal influence of the previous *corner entry* phase. In this regard, it is very useful to consider the *corner entry* as a step steer with the steering wheel angle ϕ, and use the same tools to modulate the peaks and delays. As with the step steer, the yaw rate reaches its peak and then stabilizes allowing lateral acceleration to be maximized. The duration of the yaw oscillation has an important weight in what we will call *response time* which represents the time required for the transitory phase to be completed. We can affirm that the *response time* is the sum of the time relative to the *corner entry* (ΔT of the step steer ϕ dealt with in chapter 8) and to the second part which includes the subsequent corrections to arrive at $\Delta\phi$. For this reason, the trend of the sideslip angle variation is a very significant factor in these phases and all the related actions discussed in chapter 8 (paragraphs 8.4.1 and 8.4.2) represent the most suitable tools both for adjusting the *response time* and to modulate the trend of the vehicle reaction as a function of the frequency of application of the steering wheel angle ϕ. This last adjustment is very important in order not to have the unwanted yaw response in two times which gives the vehicle a difference in behavior depending on the speed of application of the steering wheel angle and consequently a different perception of the $\Delta\phi$ to be applied to stay in trajectory.

It shall be stressed again that the transient, immediately following the application of the steering wheel angle ϕ, overlaps with the *corner entry*. The separate management serves only to divide the phenomena in order to facilitate the explanation of the vehicle reactions.

The new configuration assumed by the suspension, with the external wheels bumped (compressed) and loaded laterally, which generated both a progressive variation of the slip stiffness of the axles and a different orientation of the wheels, leads to a $\Delta\phi$ correction on the steering wheel angle. The correction, necessary to continue the trajectory with the desired radius of curvature, must not be discordant because it would imply an unstable and oscillatory behavior. The $\Delta\phi$ value must be very low in sports cars both in terms of value and *response time*.

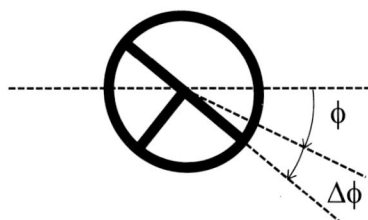

Figure 9 – Correction needed to maintain the trajectory.

The factors on which to act to vary the value of the correction, its sign and the *response time* are:

- *Camber* gain and *camber* loss under lateral load.
  As shown in chapter 6, a greater gain of *camber* in bump travel (compressed springs) corresponds to a smaller loss in roll and consequently to a smaller decrease in the slip stiffness. Since the goal of a $\Delta\phi$ concordant corresponds to a lower loss of *camber* at the rear compared to the front (when cornering), it is advisable that the rear *camber* gain in bump travel (compressed springs) is greater than the front gain and the rear *camber* loss under lateral load is smaller than the front one. When adjusting the $\Delta\phi$, which depends on the main purpose of the vehicle, in addition to properly positioning the front and rear ICFV,y must also take into account the *Suspension Jacking* phenomenon, fully described in chapter 6. The *camber* also adjusts the yaw peak and the *response time*. In particular, the *response time* increases if the leaning delay due to a high yaw peak increases, that is, if the *TR-PP(AY-r)* increases. This is lowered by increasing the rear *camber* gain and decreasing the loss under lateral load. Where a sporty behavior suitable for high speeds is required, it is necessary to boost the total increase of the rear *camber* (more *camber* gain in bump travel + less loss under load) compared to the front one, and that the two increases are formalized in the shortest possible time. The stabilization speed, which is guaranteed above all when the stiffnesses in opposite wheel travel and under load of the suspension are high, considerably lowers the response time. The above is fully described in chapter 8.

- *Toe* variation in bump travel and under lateral load.
  As seen in chapter 6, the front *toe-out* (divergence) in bump travel and under lateral load and the rear *toe-in* (convergence) in the same maneuvers contribute to increase the $\Delta\phi$. In addition, the rear closure, especially the one under load as described in chapter 8, lowers the *response time* while the front opening tends to increase it. In sports cars, in order to lower the *response time*, it is advisable to maintain a good level of rear closure (more that under load than that in bump travel) and to limit the front opening to a minimum. In road cars it is advisable to increase the front opening in order to ensure greater stability and less steering responsiveness. It shall be remembered that, even if considered a source of delay, the front divergence also contributes to increasing the *AYLIN* of the understeer curve and to imposing to the vehicle a more predictable behavior as lateral acceleration increases, as seen in chapter 8.

- Static *toe*.
  The static *toe-out* on the front axle, although it makes the vehicle more reactive in the initial part of the curve, increases the $\Delta\phi$. The static *toe-in* on the rear axle produces the same effect with regard to $\Delta\phi$. The static *toe* also affects the trend over time of the $\Delta\phi$ but the influence is less significant than the overexposed variations. In particular, the rear static *toe-in* lowers the *response time* because it decreases the ΔT.

- Roll stiffness.
  Increasing the roll stiffness of the vehicle decreases the $\Delta\phi$ both as a value and as a *response time*. This is because the vehicle finishes the roll motion faster as the sus-

pension makes less travel. The distribution of the roll stiffness affects the distribution of the slip angles of the axles. Adding roll stiffness on an axle increases the corresponding slip angle. Consequently, always referring to the step steer, as the stiffness distribution on the rear axle increases, the $\Delta\phi$ decreases but the *response time* could increase due to the greater in ΔT due to the increase in the sideslip angle.

- Longitudinal arrangement of the masses
The distribution of lateral forces on the two axles follows the criterion of the simply supported beam, and the axle most stressed laterally is the one closest to the center of gravity. This means that, with the same tires on the axles, the distribution of the slip angles follows the weight distribution. Consequently, the displacement of the masses towards the rear determines a decrease in the steering wheel angle thanks to the increase in the slip of the rear axle and the sideslip angle. In rear-wheel drive sports cars, the distribution of the masses shifted towards the rear, necessary to increase the traction capacity and to balance the exploitation of the axles, is possible as long as the following difference is positive, present in the formulation of the steering wheel angle (chapter 4).

$$\left(\frac{M_f}{C_{\alpha,f}} - \frac{M_r}{C_{\alpha,r}}\right)$$

If the latter becomes negative, because the rear mass M_r is too high, it is necessary, to avoid oversteer of the vehicle, to increase the $C_{\alpha,r}$ by increasing the width of the rear tires or by increasing the wheel closure. It shall be remembered that to increase the $C_{\alpha,r}$ it is necessary to act only on the width of the tire and not on the height (sidewall of the tire) because an increase in height (with the same under load radius) causes a decrease in stiffness. To summarize what was said in the previous paragraph, we can conclude that moving the mass distribution towards the rear increases ϕ, decreases $\Delta\phi$ and the total steering wheel angle is lowered.

- Shock absorbers
The shock absorbers act on the *response time* and on the fact that the $\Delta\phi$ must absolutely avoid having a vibratory behavior during the transient. If the shock absorbers are not properly calibrated, they produce the oscillation of the suspension with consequent variability on the $\Delta\phi$ and of the *response time*. The *response time* increases even if the shock absorbers are too braked because the vehicle is slow to reach the final roll angle. Correctly calibrated shock absorbers also help to reduce yaw oscillations after maximum peak, the absence of which lowers the *response time*. A slightly more braked adjustment in rebound travel allows to work with on average higher external *camber* (chapter 6), increasing the car reactivity and lowering the *response time*. The increase in rebound braking shall not be overdone, otherwise the apparent improvement in reactivity is overshadowed by the expansion of the *response time*, which inevitably increases because the vehicle delays in reaching the stabilized roll angle.

Automotive suspension

The $\Delta\phi$ added to the steering angle of the *corner entry* phase helps to define the *Kus*. It shall be reminded that the first two phases represent the sum of a step steer ϕ plus the necessary correction $\Delta\phi$ to stay on the trajectory.

Stabilized

This phase begins when all the quantities are stationary and the vehicle is cornering at a constant steering wheel angle. In this phase the vehicle reaches the maximum lateral acceleration (except for the peak which is not relevant for the evaluation of the final radius of curvature). Once the trajectory is fixed, the maximum velocity value relative to the maximum lateral acceleration follows the relationship approximated (because it considers the yaw rate equal to V_x/R and not to V/R):

$$V_x^2 = R(A_{Gy})$$

This is only valid in stationary conditions, in which R is the radius of curvature near the apex which is the minimum stationary resulting from the step steer ϕ and the subsequent correction $\Delta\phi$. Knowing the trajectory and the maximum lateral acceleration that can be reached by the vehicle in stationary conditions, the maximum travel speed of the curve can be calculated. If the vehicle has a high *response time*, it risks reaching the apex in unstabilized conditions. In this condition the radius of curvature at the apex, which is not the minimum relative to the step steer and subsequent correction applied, does not remain stable but tends to decrease to the stabilized value and this is an undesirable behavior because it requires a further correction to remain in trajectory. Furthermore, this leads to an inevitable lowering of the travel speed of the curve, since reaching the stationary condition also allows you to accelerate. The possibility to anticipate the longitudinal acceleration is what makes a vehicle more performing. Furthermore, a vehicle with a low *response time*, since it is able to anticipate stationary conditions, allows to brake later and to increase the *corner entry* velocity in order to increase the speed of the entire maneuver.

Once the stationary condition has been reached, it is appropriate that, in the presence of an external disturbance, the vehicle remains stable or requires the user to give a correction $\Delta\phi$ concordant with the stationary steering wheel angle. Referring to the wheels outside the curve, which influence the dynamic behavior of the vehicle the most, we shall consider when the front external wheel bumps and acquires a longitudinal stress due to an asymmetrical road hump. If the wheel opens (*toe-out*) both in bump travel and under longitudinal load, the car widens the radius of curvature and the user intuitively reacts with a concordant $\Delta\phi$. The unwanted situation occurs when, due to the wheel closing (*toe-in*) in the presence of the road hump, the user has to react with a counter-steering which represents a less intuitive maneuver than the concordant $\Delta\phi$. The same observations apply if the hump is faced by the rear wheel which must close (*toe-in*) to give stability to the vehicle. The same behavior ($\Delta\phi$ concordant) is required in the presence of an increase in driving torque.

Yaw stiffness also plays a key role in cornering stability, which brings the vehicle back on track in the face of an external perturbation. Therefore, the lateral elasticities of the suspensions that cause vehicle response delays are essential for cornering stability.

Corner exit

In this phase, the user must put the vehicle back in a straight direction in order to exploit the maximum adherence for longitudinal acceleration. How this happens depends on how the trajectory of the curve has been set. If it has been addressed correctly, the user must accelerate and release the steering wheel progressively in order to increase the radius of curvature to return to the straight. The natural understeer tendency, caused by the increase in lateral acceleration resulting from the increase in speed, tends to move the vehicle towards the outside of the curve, as shown in the following figure. This forces the driver, who wants to increase the speed of the vehicle, to delay the release of the steering wheel or to increase its angle when the lateral displacement is too pronounced. This phenomenon, if very accentuated, slows down the car because the acceleration phase takes place with the wheels still steered, that is, in conditions that are not optimal for exploiting the full longitudinal adherence of the tires (for a front-wheel drive vehicle).

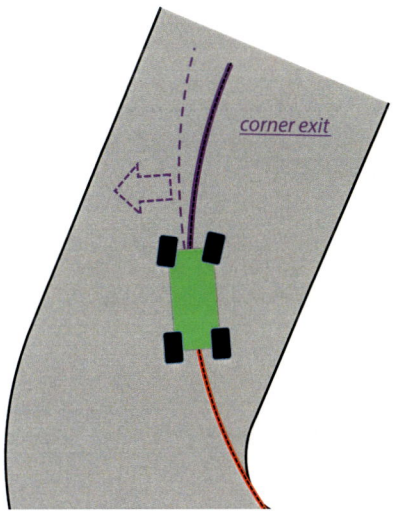

Figure 10 – Lateral movement of the vehicle due to the increase in longitudinal speed.

In order to improve this behavior, it is necessary to:
- Limit the front wheel opening both under load and in bump travel, to decrease the steering wheel angle request as the lateral acceleration increases. Always keep in mind that the cancellation of the front opening tends to lower *AYLIN*.
- Adjust the *camber* distributions between the axles (gain in bump travel and loss under lateral load) to lower the understeer.
- Move the roll stiffness distribution to the rear. This possibility of adjustment requires that the center of gravity is not too close to the roll axis, otherwise the distribution of stiffness becomes negligible in the distribution of load transfers between

the axles (chapter 3).

- Increase the rebound damping of the front shock absorbers to avoid the nose lifting during acceleration and the consequent loss of *camber* and caster. In this way, in the initial phase of acceleration, the pitch of the car is formalized with a rotation around an axis parallel to the junction between the centers of the front contact patch, favoring the caster increasing and the rear-bump travel, as shown in chapter 6.
- Lower the moment of inertia J with respect to the z axis. As seen at the end of chapter 4, the presence of a positive angular acceleration produces an increase in the steering wheel angle directly proportional to the moment of inertia J and inversely proportional to the wheelbase. Lowering J the request for lateral force on the front axle (very stressed in this phase) is limited, also attenuating the consequent effect of divergence under load. Furthermore, this stage of the curve is characterized by the overlap between the growth of lateral acceleration and the growth of angular acceleration. Lateral acceleration produces an increase in lateral forces on the two axles, while angular acceleration produces an increase in the front and a decrease in the rear of the same forces. This last decrease leads to a lowering of the slip angle on the rear axle and the consequent request for an increase in the steering wheel angle. As the slip stiffness of the rear axle increases, the lowering of the rear slip angle decreases, and the request for the steering wheel angle is reduced. In this way the exit of the curve is easier. To guarantee the truthfulness of this statement, the contributions of lateral forces must be quantified because the effect of angular acceleration (reduction of the force on the rear axle) could be overshadowed by the effect of lateral acceleration thanks to which the request of steering wheel angle always increases as the slip stiffness of the rear axle increases. If the latter effect is more important, it is advisable to lower the rear slip stiffness.

Obviously, the above actions can compromise the behavior of the vehicle in other maneuvers. It is the designer's task to balance the needs, taking into account that this phase has a heavy weight in the time spent on the curve.

The tendency to increase the radius of curvature in the presence of an increase in longitudinal speed, which precludes the possibility of accelerating fully out of corners, is a characteristic of all vehicles and increases with the increase in the understeer gradient. Front-engine and front-wheel drive vehicles suffer more from this trend because traction lowers the lateral adherence capital favoring the increase in slip at the front. In general, the improvement of this phase lowers the understeer gradient and leads to a greater reactivity of the vehicle in changes of direction which is not always appreciated due to the greater difficulty in controlling the vehicle itself.

Vehicle dynamics – Achievement of the targets

We shall also examine the situation in which a trajectory different from the optimal one is set, such as when the apex is anticipated too much and the radius of curvature is increased, as shown in the following figure.

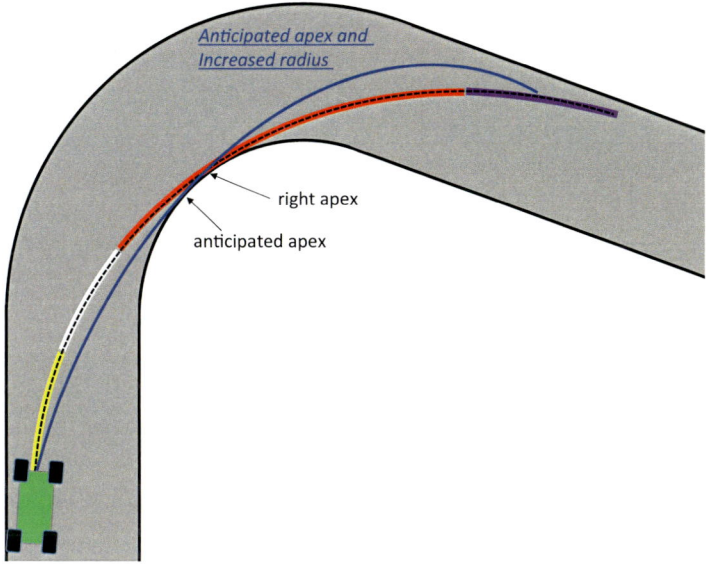

Figure 11 – Trajectory different from the optimal one.

In this case, the user must increase the steering wheel angle in order not to go off the road. If the vehicle works in the linear area of the understeer curve, it is easier to control because the increase in the necessary steering angle remains proportional to the required delta acceleration. It is clear that cars with higher *AYLIN* are easier to manage in the cornering exit, both for corrections performed due to incorrect trajectories and because they allow a faster steering wheel release.

9.3. Slalom

The ability of the vehicle to carry out this maneuver is important both for sports vehicles and for those intended for the transport of people. This is because it represents an obstacle passage with subsequent re-entry into the trajectory. In sporting applications, in addition to ensuring that the vehicle does not go into crisis, which is formalized with roll over or loss of adherence as shown in chapter 3, travel time is fundamental. We shall focus our attention on this last performance and we observe that, in order for the vehicle to be fast in the change of direction, it is necessary that, with the same slip angle imposed by the steering, the lateral force emerging on the front axle quickly reaches the maximum in order to generate the yawing moment it takes for the vehicle to turn. The maximum lateral force, with the same steering angle, increases as the slip stiffness of the front axle increases. For the yaw motion to be fast, the front lateral force must be matched by an equally sudden rear lateral force that allows the rapid achievement of the condition of maximum lateral acceleration. The lateral elasticity and damping of the rear suspension cause the rear reaction delay and a rotation around the center of gravity (sideslip angle) which slows the lateral movement of the vehicle. Below is the schematization of the slalom maneuver in phases.

The first three phases are almost the same as the first three of the curve maneuver, with the difference that the variation of the quantities is much faster and some phases disappear because they are overshadowed by the following ones.

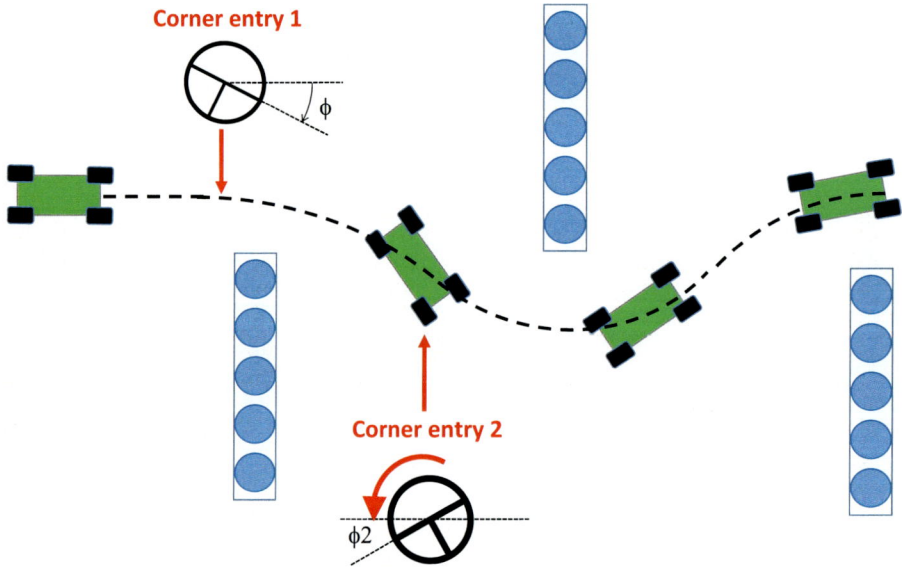

Figure 12 – Schematic of a slalom with steering wheel seen by the user.

Corner entry 1

It represents the time interval in which the user, starting from a straight running condition, applies the steering wheel angle they deem necessary to change direction. This does not differ from the *corner entry* of the curve maneuver described above. In this type of maneuver it is preferable to have *pro-Ackermann* steering which guarantees greater yaw moment and faster cornering, always taking into account the observations made in chapter 7 regarding the shift of the *sliding limit* on the inner wheel.

Transitory 1

It represents the time interval in which the vehicle completes the roll travel and the user corrects the steering wheel angle to stay on the trajectory. In this phase, the same considerations made in the transitory of the cornering in relation to the *response time* apply. When the slalom is very narrow (low radius of curvature) or takes place at high speed (as in some chicanes), the vehicle does not have time to complete all the roll travel and correction $\Delta\phi$ is applied either only partially or not at all. In this case, the behavior of the whole *corner entry* and the transient is even closer to a step steer maneuver at constant speed, and the response time consists only of the ΔT. Given the greater analogy with the step steer, it is even more evident that a vehicle with a lot of sideslip angle variability, which delays yaw stabilization and lateral acceleration growth, is more difficult to manage due to the response variability, which manifests itself above all with the need to correct the steering wheel angle compared to a vehicle that gains less sideslip angle (see chapter 8).

The low variability of the sideslip angle, which in most cases is linked to a contained gain, produces behavioral characteristics that make the car less "nervous" in the change of direction which consist in:

- Fewer vibratory reactions at the steering wheel;
- More constant radius of curvature during the transitory;
- Greater radius of curvature with the same steering wheel angle;
- Lower *response time*. In chapter 8 we see that also with the same lateral acceleration, obtained through the steering wheel angle increase, the car with the lower sideslip gain has a more contained ΔT.

Considering this we can affirm that, in the changes of direction at high speed, a lower sideslip angle gain is certainly indicated because it corresponds to a more stable behavior. At low speed, on the other hand, a greater sideslip angle gain could be effective. This, with the same steering wheel angle, lowers the radius of curvature and makes the car more agile at the price of greater instability.

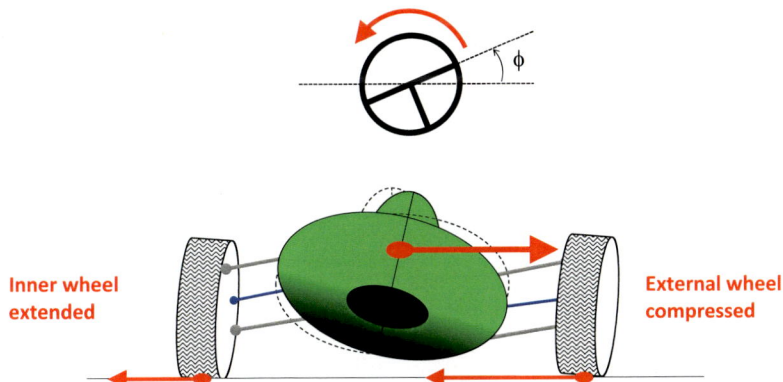

Figure 13 – Front view of the vehicle engaged in a right cornering at the end of transitory 1 showing the loss of *camber* on the outer wheel and on the inner wheel. The steering wheel is shown in the same view as the vehicle.

Another important consideration concerns the calibration of the extension of the shock absorbers, given the speed of carrying out this phase. As explained in the curve transitory and in chapter 6, the increase in rebound damping produces more favorable average *camber* values because it causes the outer wheels to lose *camber* more slowly. Obviously, everything must be balanced between the axles because if, for example, the extension characteristic is increased only on the front axle, there is a risk of oversteer as the front external *camber* is on average higher. If, on the other hand, only the rear extension characteristic is increased, there is a risk of excessive understeer which is formalized in the lower cornering capacity.

Stabilized 1

It represents the time interval in which the car curves with the stabilized quantities. This phase is non-existent when the slalom is sudden.

Corner entry 2

It represents the time interval in which the user applies the steering wheel angle $\phi 2$ (in the opposite direction to the initial *corner entry*) that they deem necessary to follow the imposed trajectory. For simplicity we will consider the angle value $\phi 2$ starting from the straight steering wheel configuration. In this sudden change of direction, which occurs with the car still rolled, the steering of the wheels is much faster than the consequent roll in the opposite direction. For this reason, when the car begins to turn, as controlled by the user, it finds itself in the configuration shown in the following figure

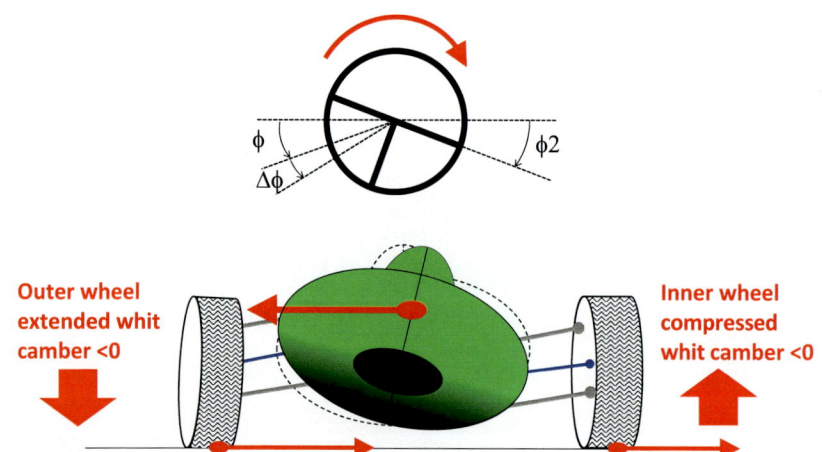

Figure 14 – Front view of the vehicle controlled to turn left in corner entry 2. Since the car still has the roll angle of the previous phase, the outer wheel is still extended and has negative *camber* with respect to the curve. The steering wheel is shown in the same view as the vehicle.

The main features of this configuration are the following:
1. The outer wheels are still extended and have apparently favorable *camber* angles because they are negative with respect to the direction of the curve. The configuration of the vehicle is only apparently favorable as the *camber* angles

outside the curve are not well balanced between the axles. This is because the optimal sizing of *camber* variations (see chapter 6) requires the rear axle to have less loss in roll with respect to the front axle. For small roll angles, the *camber* variation on the two wheels of the same axle is the same, and consequently the inner rear wheel loses less *camber* than the inner front wheel. When the vehicle suddenly changes direction, the loss of *camber* of the inner wheels becomes gain. Consequently, the rear wheel, having lost less in the previous direction, has less gain than the outer front wheel. Therefore, in the first part of *corner entry* 2 the vehicle turns with less *camber* gain at the rear which makes the car even more direct. This situation, if aggravated by an unfavorable combination with static *camber* and a sub optimal *camber* gain distribution (in bump-rebound travel), could lead to oversteer because the decrease in lateral stiffness, due to the loss of *camber*, could become greater at the rear. The following figure shows the comparison between the configurations at the end of the stabilized section 1 and at the beginning of the *corner entry* 2 in which the external wheels are still extended. The representation does not consider the static values of *camber* but focuses only on the differences between the roll losses.

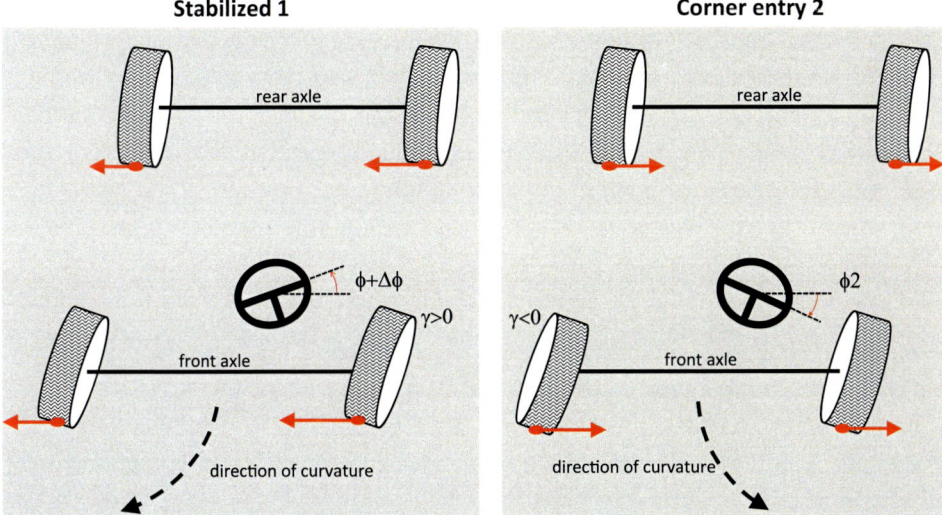

Figure 15 – Comparison, in three-dimensional view, between the combinations of the *camber* angles between the front axle and the rear axle in the configurations relative to the end of the stabilized section 1 and the beginning of the corner entry 2. The *camber* angles of the beginning of the corner entry 2 are negative with respect to the new direction of curvature but in an unfavorable condition because the front *camber* exceeds the rear one.

As anticipated, the situation worsens when the difference between the rear *camber* gain and the front *camber* gain (in bump-rebound travel) increases, which corresponds, as shown in chapter 6, to the difference between the roll losses. For this reason, it is important not to overdo this difference too much.

2. When considering the *toe,* the extended outer front wheel is closed while the rear one is open. This kinematic configuration induces oversteer.
3. The outer wheels can be further extended due to the lifting component of the *Suspension jacking* (chapter 3) caused by the excessive height of the roll center. Since this effect emphasizes several aspects already described, a car that has to deal with this maneuver must not have the roll axis too high.
4. The outer front wheel has less *camber* gain in steering as it has lost caster in extension. When this phenomenon is very pronounced, the lower *camber* gain and the low vertical load on the outer wheel could significantly lower the yawing moment produced by the outer wheel and delay or even not allow the sudden change of direction.
5. The limited vertical load of the external front wheel considerably lowers the characteristic curve of the tire which cannot guarantee adequate lateral force values to the slip imposed by the steering. On the other hand, the internal front wheel, having greater vertical load, in addition to guaranteeing greater lateral force to slip imposed also has the *sliding limit* still high. Therefore, to improve this phase, it is possible to think of increasing the *Ackermann* percentage to boost the yawing capacity of the vehicle without causing a crisis in the inner tire. In chapter 7 a more detailed explanation of the phenomenon.
6. The extended configuration of the outer wheels produces an increase in the loss of *camber* under lateral load caused by the increase in the moment of force due to the greater distance of the ground point from the suspension mounts.

   The external wheels still unloaded have undergone a decrease in the slip stiffness which lowers, and sometimes reverses, the difference between the lateral forces of the axle since, as seen in the last paragraph of chapter 2, the distribution of these forces follows the distribution of the slip stiffness. This means that the wheels outside the curve either have little increment or have less lateral force than the inside ones and this is positive because it reduces loss of *camber* under load.

In these conditions the vehicle risks the crisis that can be formalized in oversteer when the points 1,2,3 have greater impact and in understeer when point 4 has a lot of impact. The conditions of the vehicle in *corner entry* 2 are deliberately generated in the Scandinavian pendulum technique which is used in rallies. This consists in the application of an initial step steer opposite the direction of the curve. In this way the vehicle faces the curve with the external wheels unloaded which, together with the transfer of load on the front axle due to the release of the gas pedal, produce a controlled oversteer. This controlled skid technique is used in low-adherence situations to avoid understeer and allows you to line up with the next straight before you have completed the corner, at this point having the opportunity to straighten the front wheels and accelerate hard much earlier compared to the normal travel. In general, controlled skid is a disadvantageous technique when there is an excellent adherence due to high-performance tires, good asphalt and suspensions that do not induce understeer.

The configuration of *corner entry* 2 also occurs when, in the face of oversteer of the vehicle, excessive counter-steering is performed which can degenerate into oversteer in the opposite direction to the previous one.

For all the reasons mentioned, vehicles with large suspension travel are not suitable for slalom maneuvers due to both the excessive variation in *camber* and *toe* and the delay in return due to the large excursion.

Vehicles designed for high speeds, which to favor understeer and lower the sideslip angle, have a greater difference between the front and rear *camber* gain, suffer from the aforementioned imbalance between the internal gains which, together with the *toe* variation, always dimensioned in understeer optics, can cause the vehicle to oversteer.

Transitory 2

Time interval in which the roll excursion is completed and the consequent correction on the steering wheel takes place. This phase differs from transitory 1 due to the greater roll excursion since when the steering wheel angle is applied the car is rolled to the opposite side.

Return in trajectory

It represents the interval in which the user applies the steering wheel angle that allows them to re-enter the trajectory. There are no specific critical issues during this phase.

In conclusion, we can say that to face the slalom maneuver quickly, the vehicle must have the following characteristics:

- Limited roll and low suspension travel;

- Low roll axis to limit *suspension jacking*;

- Small difference between the rear *camber* gain and the front *camber* gain in bump travel. It is advisable to reduce this difference to the minimum limit ($\Delta\phi$ positive), increasing the gain at the front, to make the *corner entry* 2 more stable. The increase in the front *camber* gain also helps the corner exit phase of the cornering maneuver.

- Limited front opening (*toe-out*) in bump travel because in *corner entry* 2 it turns into closure (*toe-in*) for the outer wheel. If front *toe-out* is necessary, it is advisable to increase the *toe-out* under load.

- Limited rear closing (*toe-in*) in bump travel because in *corner entry* 2 it turns into opening (*toe-out*) for the outer wheel. If rear *toe-in* is required, it is advisable to increase that under lateral load both in order not to bring the vehicle close to the oversteer and in order not to generate too much variability of the sideslip angle as shown in paragraph 8.4.2 of chapter 8.

- Shock absorbers with a slight increase in the extension characteristics because, in addition to limiting the rebound of the external wheels, they improve the configuration of the vehicle in the first part of the maneuver, as explained in chapter 3 and in the transitory 1.

- Higher *Ackermann* percentage. To optimize this maneuver, the *Ackermann* percentage can be increased both to have greater agility and to improve *corner entry* 2 as described above.

It shall be stressed again that these characteristics are less suitable in vehicles intended for fast routes because they give a reactivity that is difficult to manage at high speed.

Analytical index

Some terms listed in the index have been repeated to facilitate the search.
In some cases the pages associated with them contain the topic but not the term.

A

Acceleration
    angular, 92, 93, 94, 95, 96, 97, 236, 242
    centripetal, 22, 58, 62, 95.
    center of gravity, 82.
    lateral, 22, 59, 60, 64, 67, 69, 77, 82, 87, 89, 90, 92, 96, 97, 98, 136, 177, 187, 191, 199, 205, 207, 214, 215, 224, 234, 235, 236, 237, 238, 241, 242, 244, 245.
    lateral at constant velocity, 83, 210, 211, 212, 214, 215, 216, 217, 218.
    longitudinal 12, 161, 227, 230, 231, 240, 241.
    maximum lateral 59, 60, 91, 240.
    peak 211, 216, 220.
    sprung mass, 73.
    stabilized, 215.
Accelerometer, 77.
Ackermann 124, 133, 186, 189, 191, 192, 244.
    error 187, 188, 189.
    percentage, 124, 133,188, 191, 192, 206, 234, 248, 250, 250.
Adhesion area, 23, 24, 26, 27, 28.
Adherence
    capital, 34, 36, 130, 143, 162, 187, 242.
    coefficient, 33, 34, 35, 36, 40,59.
    ellipse, 34, 231, 232.
Adhesion length, 27,28.
Agility, 96, 188, 191, 235, 250.
Anti dive, 126, 148, 161, 162, 163, 165, 166, 168, 169.

Anti lift, 148, 163, 164, 165, 166, 168, 169, 170.
Anti-roll bar, 16, 52, 54, 55, 56, 57, 64, 68.
    external diameter, 57.
    internal diameter, 57.
    lambda bar, 55, 56.
    sizing, 55.
    torsional stiffness, 56, 57.
Anti squat, 148, 163, 164, 165, 166, 170.
Apex, 12, 13, 233, 234, 240, 243.
    anticipated, 243.

B

Ball joint, 55, 100, 101, 106, 108, 109, 110, 130, 136, 138, 142, 145, 148, 149,167, 171, 172, 195, 196.
Bicycle (equivalent), 84,199.
Bilink, 111, 115, 116.
Bourlet, 192.
Braking, 12, 14, 33, 34, 35, 36, 38, 42, 60, 75, 124, 125, 130, 133, 142, 143, 161, 162, 163, 168, 169, 170, 183, 185, 188, 227, 229, 230, 231, 232, 233, 239.
    asimmetryc, 185, 231, 232.
Brush model, 23, 24.
Bump travel
    frequency, 46, 48.
    stiffness, 46, 55.
Bump steer, 135, 138, 139, 195, 196.
Bump stop, 70, 107, 205.
    clearance,70, 207.
    lambda,70.

251

Bushing (elastic), 100, 101, 102, 103,104, 106, 112, 113, 114, 116, 118,119.
    external sleeve, 102.
    internal sleeve, 102.
    stiffness, 102, 103, 104.

C

Calculation code, 104, 114, 197, 198.
    FEM, 57, 113, 114.
    Multibody, 79, 96, 197, 198.

Camber
    equilibrium between axles, 126, 127, 128, 129, 130, 156.
    effect on the characteristic curve, 30,190.
    gain,17, 32, 105, 109, 113, 114, 116, 118, 119, 121, 128, 129, 148, 152, 154, 155, 156,157, 159, 173, 174, 175, 183, 189, 190, 196, 202, 206, 207, 220, 225, 235, 238, 247, 248, 249.
    loss under lateral load, 101, 105, 109, 125, 129 130, 202, 203, 206, 207,214, 238, 241, 248.
    static, 38, 42, 126, 127, 206, 207, 214, 224, 228, 234, 235, 247.
    tolerances, 233.
    variation in roll/curve, 153, 154, 155,156, 157.

Carcass, 23, 30, 31, 32, 40.

Caster, 125, 132, 167, 170, 171, 172, 173, 174, 175, 176, 178, 179, 206, 227, 228, 229, 230, 233, 235, 236, 242, 248.
    first stabilizing action, 173, 175.
    measure, 174.
    rear negative,115, 132.
    second stabilizing action, 175, 178, 228.
    variation, 167, 170, 174.

Caster trail, 125, 172,173, 176, 177, 179,181, 228, 229.

Center of gravity
    acceleration, 82, 200.
    height, 16, 59, 60, 149, 205.
    longitudinal position, 17.
    of the vehicle, 149, 198, 208, 236.
    of the sprung mass, 61.
    rotation around the, 208, 213, 215, 221, 244.
    velocity, 80, 82, 213.

Center of instant rotation, 84, 145, 186, 187.
    construction, 151
    measure 158,159

Centrifugal force, 16, 20, 41, 52, 58, 59, 61, 85, 94, 124, 157, 205.

Change of direction,14, 22, 136, 175, 209, 211, 234, 243, 245, 246, 248.

Characteristic angles of the wheels, 43, 106, 109, 117, 124, 125, 126, 128, 144, 162, 170, 189, 197.

Characteristic of the tire (lateral), 33, 35, 36, 39, 138, 140,141,235.
    central section, 21, 22.
    final section, 21, 22.
    initial section, 21, 22.
    of the axle, 41, 93, 127, 138,139,140,141, 200, 201, 202, 203, 206, 207.

Coefficient
    of adherence, 33, 34, 35, 36, 59.
    of friction, 24, 26, 27, 34, 40.

Comfort, 45, 52, 72, 73, 74, 76, 99, 109, 114, 123, 130, 163, 164, 168, 169, 170.

Cones, 14, 221.

Contact patch, 22, 23, 24, 25, 26, 27, 31, 32, 33, 36, 37, 39, 75, 150, 151, 159, 168, 172, 174.

Control arm, 43, 55, 100, 102, 103, 108, 146, 146, 167.

Cornering, 12,233
    corner entry,13, 234.
    corner exit, 14,241.
    transitory, 13, 236.
    stabilized 13,240.

Correction of the steering wheel angle, 13,237.

Counter-steer, 13, 90, 128, 142, 202, 241, 240, 249.

D

Damping
- critical, 74.
- optimal for comfort, 73, 74.
- optimal for handling, 73, 74.

Differential, 181.

Distribution of the slip angles, 17, 64, 76, 239.

Distribution of the weight, 63, 66, 68, 76, 77, 96, 201, 236, 239.
- effect on corner entry and on transitory, 96, 236, 239.
- effect on understeer, 201, 236, 239.

Drag force, 134, 191.

Dynamic friction force, 27.

E

Elastokinematics, 99, 100, 103, 105, 124, 191, 203, 224, 225, 234.

Equivalent link, 146, 147.

Euler, 197.

Experimental approach, 79, 83.

External work, 47.

F

Fork phasing angle, 193, 194.

Friction, 24, 25, 26, 27, 33, 34, 40, 60, 177, 183, 232.
- coefficient, 24, 26, 27, 34.

G

Geometric quantities of the suspension, 124, 125, 126, 171.

Gyroscope, 77, 83, 198.

H

Handling, 71, 72, 76, 99, 123, 197.

Homokinetic, 193, 194, 195.

Homokinetics, 193.

I

ICFV - Transversal instant center, 65, 143, 144, 145, 146, 148, 151, 152, 153, 154, 155, 156, 157, 158, 159, 238.
- construction, 145, 146.
- position calculation, 158, 159.

ICSV - Longitudinal instant center, 143, 144, 145, 147, 148, 160, 161, 162, 163, 166, 167, 168, 170.
- construction, 147.
- position calculation, 158, 159.

Inertial tensor, 92.

Instability, 90, 112, 179, 220, 227, 235, 245.

Instantaneous rotation axis, 113, 116, 117, 143, 144, 145, 147, 148, 149, 196.
- construction, 145, 146, 147.

Internal work, 47.

J

Jacking, 66, 149, 157, 238, 248, 249.

Janteaud, 192.

Joint
- ball joint, 55, 100, 101, 106, 108, 109, 110, 130, 136, 138, 142, 145, 148, 149, 167, 171, 172, 195, 196.
- cardan, 193.

K

King Pin Inclination, 6, 106, 125, 179, 228.

KPI_{offset}, 125, 180, 181, 182, 183, 184, 185, 230, 231.

L

Lagrange, 197.

Lambda bar $(Lamba,b)$, 55, 56.

Lambda bump stop $(Lambda,bump)$, 70.

Lambda shock absorber $(Lambda)$, 73, 74.

Lambda spring $(Lambda)$, 46, 47, 48, 49, 50, 69, 73, 107.

Lateral force, 41, 62, 63, 64, 73, 94, 103, 104, 112, 114, 115, 124, 125, 134, 136, 139, 140, 141, 176, 177, 216, 219, 220, 221, 236, 242, 244.
- surplus, 96, 98, 220.

Lateral leaning, 136, 211, 238.

Lifting moment, 65, 66, 157, 158.

Load transfer, 61.
- contribution of the lateral force, 63.
- contribution of the roll, 63.
- contribution of the unsprung masses, 63.
- contribution of the weight force, 64.

Longitudinal

Force, 35, 124, 125, 134, 161, 180, 181, 185, 191, 230, 231, 232.

Kinematics, 147,161.
- load at ground, 124.
- load at wheel center, 124.

Loss of camber, 38, 66, 101, 105, 109, 125, 129, 129, 152, 153, 154, 155, 156, 157, 158, 160, 179, 202, 203, 206, 214, 238, 242, 245, 247, 248.
- on roll, 153, 154, 155, 156, 157.
- under load, 101, 109, 129, 203, 206, 238, 248.

M

Main axis of inertia, 149, 150.

McPherson, 17, 70, 105, 108, 109, 110, 111, 114, 115, 145, 147, 148, 149, 167, 172, 182, 183, 184, 185, 193, 202, 203.

McPherson with revolving upright, 110, 111.

Misalignment of the steering wheel, 130, 131, 132, 227, 230, 233.

Moment of inertia, 45, 57, 64, 92, 96, 98, 107, 220, 236, 242.

Motion
- bump travel, 45.
- pitch, 45.
- plane, 19, 79, 84, 150, 208.
- rebound travel, 45.
- roll, 52.

Multibody, 79, 96, 197, 198.

Multilink, 111, 114, 116, 117.

N

No stationary condition, 92.

O

Obstacle passage 14, 15, 72, 125, 142, 168, 169, 180, 243.

Opposite wheel travel
- motion, 124,128
- stiffness, 54, 55, 58, 205, 239.

Oversteer, 17, 64, 90, 91,94, 96, 98, 112, 114, 121, 142, 157, 174, 191, 201, 204, 235, 239, 246, 247, 248, 249.

P

Pacejka, 197.

Panhard, 192.

Pavè, 72.

Pinion and rack, 17,193,194, 195.

Pitch, 19, 45, 52, 72, 75, 76, 81, 92, 150, 161, 163, 169, 198, 242.

Pivots, 143, 147, 148.

Principle of virtual works, 40.

Pull rod, 107, 108.

Push rod, 107, 108, 236.

Q

Quadricycle, 19, 20, 79, 80, 81, 84.

R

Rack housing, 17, 103, 195, 196.
- advanced, 195.
- set back, 195.

Radius of curvature, 17, 41, 53, 84, 89, 90, 91, 142, 143, 208, 217, 218, 220, 221, 224, 225, 237, 240, 241, 242, 243, 245.
- minimum, 220.

Rally, 248.

Reactivity, 133, 201, 239, 240, 242, 250.

Rebound steer, 135.

Rebound travel
- frequency, 46, 48.

motion, 45.

stiffness, 46.

Reference system, 20, 21, 152.

Relaxation length, 6, 23, 221, 222, 225.

Resistance to advancement, 130, 132, 133, 134, 172, 173, 191, 228, 229.

Response time, 237, 238, 239, 240, 244, 245.

Rim

offset, 180, 183, 184, 185.

mating surface, 183, 184.

Rocker, 107, 108.

Roll

angle, 13, 17, 52, 60, 63, 67, 71, 77, 123, 154, 155, 156, 160, 205, 222, 223, 225, 239, 240, 246.

gradient, 67, 68, 69, 199, 205.

moment, 67.

vibration, 223.

Roll axis, 52, 54, 61, 63, 64, 66, 67, 76, 77, 149, 150, 205, 242, 248, 260.

Roll center, 63, 65, 66, 76, 126, 149, 150, 151, 152, 153, 155, 157, 158, 160, 248.

Rolling resistance, 32, 33.

Roll over, 16, 59, 60, 243.

Roll stiffness, 52, 54, 57, 59, 61, 62, 63, 64, 67, 68, 69, 76, 81, 85, 124, 205, 239, 242.

sizing, 67, 68, 69.

S

SAO (steering axis offset), 126, 179.

Scandinavian pendulum, 248.

Scrub radius, 125, 180, 183, 184, 185, 186, 188, 229, 232.

Semicorner, 46, 47, 48, 70, 72, 73, 74.

Shock absorber, 16, 17, 44, 50, 51, 55, 64, 70, 71, 72, 73, 74, 75, 76, 77, 107, 108, 109, 110, 111, 112, 115, 116, 145, 146, 147, 149, 159, 160, 172, 182, 198, 203, 222, 223, 224, 225, 239, 242, 246.

alteration of the kinematics, 159.

compression travel, 50, 51.

critical damping, 74.

dynamic influence, 75.

extension travel, 50, 51.

maximum length, 50, 51.

minimum length, 50, 51.

optimal comfort damping, 73, 74.

optimal handling damping, 73, 74.

total travel, 50.

Sideslip angle, 80, 83, 84, 86, 87, 88, 89, 98, 114, 132, 133, 162, 174, 199, 202, 204, 207, 210, 211, 212, 213, 214, 215, 216, 217, 218, 219, 220, 221, 222, 223, 224, 225, 235, 237, 239, 244, 245, 249.

gain, 220, 221, 245.

stabilization time ΔT, 213, 214, 215, 216, 217, 218, 219, 220, 221, 223, 237, 238, 239, 245.

Sideslip gradient, 87, 204.

Single-track model, 84, 85, 86, 89, 98.

Slalom, 9, 11, 14, 67, 220, 221, 235, 244, 245, 246, 249.

Sliding area, 23, 26, 27, 31.

Sliding limit, 28, 31, 32, 34, 36, 39, 42, 94, 121, 127, 130, 139, 177, 188, 189, 190, 191, 202, 203, 206, 207, 234, 244, 248.

Slip angle

angles distribution, 17, 76, 64, 76, 239.

evaluation, 80, 83.

increase, 41.

to the axle, 138, 139, 140, 141.

to the tire, 21, 138, 139, 140, 141.

Slip stiffness

tire, 21, 40, 139, 140, 141.

of the axle, 41, 85, 87, 128, 174, 175, 210, 214, 215, 216, 224, 242, 244.

with respect to the axle, 138, 139, 140, 141.

Solid angle, 193, 194.

Spin, 144, 167, 168, 170.

Spinning, 81, 94, 142.

Spring, 46, 47, 48, 49, 50, 51, 55, 56, 68, 70, 71, 73, 107, 108, 115, 116, 162.

compression travel, 50.

extension travel, 50.

free length, 48, 50.

installed length, 50.

lambda spring, 46, 47, 48, 49, 50, 55, 56, 57, 69, 70, 74, 107.

minimum length, 50, 51.

preload, 47, 48, 49, 50, 108

sizing, 17, 48.

Sprung mass, 43, 44, 45, 46, 47, 48, 50, 52, 61, 62, 67, 69, 72, 73, 74.

center of gravity, 61.

Stability, 90, 137, 138, 172, 175, 188, 230, 233, 238, 241.

Static steering torque, 177.

Stationary condition, 13, 14, 21, 39, 41, 58, 61, 67, 85, 88, 89, 95, 98, 159, 208, 240.

Steering

anti-Ackermann, 187, 188, 189, 190, 191.

axis, 105, 106, 109, 110, 114, 116, 168, 171, 172, 173, 174, 176, 178, 179, 180, 181, 182, 183, 184 ,193, 195, 196, 228, 229, 230, 231.

axis offset (SAO), 126, 179.

feeling ,11,13, 176, 177, 181, 183, 184.

kinematics, 20, 124, 171, 177, 186, 192, 193, 194, 198, 227, 228, 230, 232.

lever, 17, 192, 193, 196.

pro-Ackermann, 187, 188, 189, 191, 192, 234, 244.

ratio, 16, 17, 89, 91, 130, 131, 177, 193, 194, 195, 196, 198, 203, 206, 207, 220, 221, 224, 231, 234.

reactivity, 133, 135, 201, 242, 250.

Steering column, 193, 194, 195.

Steering pad, 87, 197, 199, 200, 205, 225.

Steering rod, 100, 101, 103, 106, 114, 142, 148, 149, 171, 195, 196.

Steering wheel

frequency, 210.

misalignment, 130, 131, 132, 227, 230, 233.

vibration,175, 177, 180.

Steering wheel angle, 13, 90, 128, 175, 201, 203, 225, 235, 236, 240, 244.

correction, 13, 15, 130, 227, 236, 237, 240, 245, 249.

initial, 13, 90, 135, 235, 236.

kinematic, 86.

request, 96, 241, 242.

Steering wheel torque, 174, 175, 177, 182, 183, 184, 185, 199, 227, 228, 229, 232.

dynamic torque, 177.

static steering torque, 177.

Step steer, 197, 208, 225.

Stiffness

in bump travel, 46, 55.

in opposite wheel travel, 54, 55, 58, 239.

Straight line driving

at constant speed, 12, 227.

at decreasing speed, 12, 231.

at increasing speed, 12, 230.

Suspension

bilink, 111, 115, 116.

dependent wheels, 44, 99, 105, 111, 112.

double wishbone, 17, 70, 100, 105, 106, 107, 108, 109, 110, 111, 114, 115, 116, 117, 146, 147, 148, 149, 171, 181, 182, 183, 185, 193, 202, 203.

independent wheels ,44, 99, 105, 111, 115.

jacking, 66, 149, 157, 238, 248, 249.

McPherson, 5, 17, 70, 105, 108, 109, 110, 111, 114, 115, 145, 147, 148, 149, 167, 172, 182, 183, 184, 185, 193, 202, 203.

McPherson with revolving, 110, 111.

Multilink, 111, 114, 116, 117.

semi-trailing arm,111,118.

trailing arm,111,118.

twist beam axle, 111, 112, 113, 114, 116, 135, 136, 215, 220.

virtual centers, 182, 185.

T

Toe, 36, 38, 105, 112, 117, 118, 120, 121, 130, 131, 132, 133, 134, 135, 136, 137, 138, 139, 140, 141, 142, 143, 144, 148, 162, 170, 190, 206, 207, 220, 223, 227, 228, 230, 231, 233, 235, 238, 248, 249.

    asymmetry front axle, 130.

    asymmetry rear axle, 132.

    distribution between the axle, 135, 136

    effect on the lateral characteristic, 138.

    in, 38, 114, 117, 130, 133, 134, 136, 137, 139, 140, 141, 148, 170, 206, 207, 214, 217, 218, 219, 220, 231, 235, 238, 240, 249.

    out, 125, 132, 133, 134, 135, 136, 137, 139, 140, 142, 143, 168, 176, 190, 206, 207, 215, 224, 228, 231, 238, 240, 242, 249.

    stabilizing effect, 136, 137.

    static opened, 133, 134, 135, 141, 207, 228, 235, 238.

    static closed, 133, 134, 135, 141, 207, 220, 235, 238.

    tollerances, 233.

    variation in bump travel, 118, 120, 135, 148, 196, 238.

    variation under lateral load, 135, 136, 238.

    variation under longitudinal load, 5, 142, 230.

Tire

    adhesion area, 23, 24, 26, 27, 28.

    adhesion coefficient, 33.

    brush model, 23, 24.

    characteristic curve, 21, 22, 29, 30, 31, 32, 33, 35, 39, 139, 140, 141, 188, 190, 248.

    compound, 33, 197, 229.

    pressure, 24, 25, 27, 33, 36, 37, 38, 39, 42, 222.

    relaxation length, 6, 23, 221, 222, 225.

    rolling resistance, 32, 33.

    shoulder, 222, 235, 239.

    sliding area, 23, 26, 27, 31.

    temperature, 3, 37, 38, 42, 130, 191.

    under load radius, 54, 63, 235, 239.

    vertical load, 39.

    vertical stiffness, 44, 46, 49, 72, 73, 74.

    width, 36, 85, 222.

Tolerances, 233.

Top mount, 107, 167.

Traction (driving force), 34, 35, 125, 163, 164, 165, 166, 168, 169, 170 181, 231.

Trajectory, 12, 13, 14, 15, 17, 80, 83, 84, 123, 128, 132, 136, 171, 174, 181, 186, 196, 199, 213, 220, 228, 232, 237, 240, 241, 243, 245, 249.

    return to, 15, 249.

Transverse modulus of elasticity, 57.

Tread, 23, 31, 32, 33, 37, 42, 130.

    band, 37, 38, 133.

    temperature, 37, 38, 42, 130, 191.

Twist beam axle, 111, 112, 113, 114, 116, 135, 136, 215, 220.

U

Under load radius, 54, 63, 235, 239.

Understeer, 4, 6, 17, 70, 76, 89, 90, 91, 92, 93, 96, 98, 127, 132, 133, 135, 142, 143, 157, 188, 191, 199, 200, 201, 202, 203, 204, 205, 206, 207, 211, 212, 221, 225, 235, 238, 241, 242, 243, 246, 248, 249.

Understeer curve, 90, 91, 98, 127, 132, 133, 135, 199, 201, 202, 203, 204, 205, 206, 207, 211, 221, 225, 238, 243.

    at constant radius, 211.

    at constant velocity, 211.

Understeer gradient, 89, 91, 92, 157, 200, 211, 212, 242.

    at constant velocity, 92, 211, 212.

Unsprung mass, 43, 44, 46, 49, 52, 58, 62, 63, 72, 107.

Upright, 43, 100, 101, 105, 106, 107, 108, 109, 110, 118, 143, 167, 171, 172.

V

Vehicle

    agility, 96, 188, 191, 235, 250

    drift of the vehicle, 227, 230, 231, 232.

more direct, 17, 91,127,128,133, 175, 196, 206,220,235,236,247.

more reactive, 160,235,238.

Velocity

    max when cornering, 240.

    of yaw, 93, 94, 208, 209, 212, 214, 215, 221, 222, 223, 224, 225, 240.

Virtual approach, 79, 197.

W

Wheel axis, 23, 24, 26, 31, 101, 183, 184.

Wheelbase

    increase, 96, 164, 236, 242.

    variation, 162, 166, 168, 169, 170.

Wheel bearing, 100, 101, 106, 109.

Wheel closure

    in bump travel, 113, 116, 121, 148, 170, 196, 207, 218, 219.

    under load, 114, 116, 132, 139, 220.

Wheel opening ,103, 114, 137, 148, 168, 176, 196, 203, 206, 207, 215, 221, 225, 230, 241.

    in bump travel, 114, 148, 168, 196, 207, 215, 241.

    under load, 112, 103, 137, 176, 203, 215, 230, 241.

Wheel rate, 49, 104.

Wheel ride rate, 49, 104.

Wheel track, 13, 14, 16, 54, 59, 60, 69, 126, 151, 152, 154, 159, 205, 207.

Wind gust, 178.

Y

Yaw

    gain, 211, 212.

    moment, 32, 92, 94, 134, 244, 248.

    oscillation,237,239.

    peak, 209, 210, 214, 219, 220, 221, 224, 238.

    speed, 81, 83, 93, 94, 98, 136, 137, 198, 208, 209, 210, 211, 212, 214, 215, 220, 221, 223, 224, 225, 237 240.

Yaw rate, *see* Yaw velocity.

List of main symbols and abbreviations

Some symbols have a double meaning

a	Front half-wheelbase, contact patch half-length
ay	Acceleration of the vehicle center of gravity
AYLIN	Length of the linear section of the understeer curve
AYLIN BETA	Length of the linear section of the sideslip variation in the Steering pad
AYMAX	Maximum lateral acceleration of the understeer curve at constant radius
$A_{Gy,max}$	Maximum lateral acceleration before the crisis
$\mathbf{A}_G$	Acceleration of the vehicle center of gravity
ABmax	Maximum distance between the attachment points of the shock absorber (fully compressed)
ABmin	Minimum distance between the attachment points of the shock absorber (fully extended)
α,l	Left slip angle
α,r	Right slip angle
α_a	Slip angle with respect the center line of the vehicle
α_t	Slip angle with respect the center line of the tire
b	Rear half-wheelbase, half-width of the contact patch
β	Sideslip angle
$\beta,stab$	Stabilized value of the sideslip angle in a step steer
β_0	Sideslip angle at zero velocity
C	Steering wheel torque, shock absorber damping
$C,Ackermann$	Instant rotation center of the vehicle with *Ackermann* steering
$C\alpha$	Slip stiffness
$Cbump,shock$	Bump damping of the shock absorber
$Ccompr$	Compression travel of the spring
$Ccrit,shock$	Critical damping on the shock absorber
$Cext$	Extension travel of the spring
$Cext,shock$	Rebound damping (extension) of the shock absorber
Cl	Slip stiffness of the left tire
$Cott$	Damping at ground that that minimizes vertical accelerations
$Ccrit$	Critical damping at ground
$Cott,shock$	Damping on the shock absorber that minimizes vertical accelerations
Cr	Slip stiffness of the right tire
Ct	Center of the tire contact patch

Automotive suspension

CT	Caster trail (longitudinal arm at ground)
cy	Distance in y between the ICFV and the Ct
cz	Distance in z between the ICFV and the Ct
χ	Ratio between the rear loss of camber and the front loss of the camber
d	Distance between the center of gravity and the roll axis
δ	Steer angle of the wheel
δ_s	Steering wheel angle resulting from the asymmetry of the semi-convergences
δ_0	Steer angle of the wheel at zero speed
δ_1	Steer angle of the inner wheel
δ_2	Steer angle of the outer wheel
ΔFz	Total load transfer
$\Delta Fz,total$	Total load transfer
$\Delta Fz,f$	Front load transfer
$\Delta Fz,r$	Rear load transfer
$\Delta Fz,roll,f$	Load transfer due to the roll torque distributed on the front axle
$\Delta Fz,lateral,f$	Load transfer due to the centrifugal force distributed on the front axle
$\Delta Fz,Mns,f$	Load transfer due to the front unsprung mass
$\Delta Fz,wf,f$	Load transfer due to the transverse displacement of the front weight force
$\Delta Fz,roll,r$	Load transfer due to the roll torque distributed on the rear axle
$\Delta Fz,lateral,r$	Load transfer due to the centrifugal force distributed on the rear axle
$\Delta Fz,Mns,r$	Load transfer due to the rear unsprung mass
$\Delta Fz,wf,r$	Load transfer due to the transverse displacement of the rear weight force
Δ	Difference between the peak of the yaw rate and the stabilized value
Δ,max	Difference between the maximum peak of the yaw rate and the stabilized value
$\Delta\gamma$	Camber gain in bump travel
$\Delta\gamma,$	Loss of camber in roll motion
ΔT	Delta tread temperature, stabilization time in the step steer
ΔY	Delta between the lateral forces acting on the two wheels of the axle
ET	Rim offset
ε	Toe angle
ϕ	Starting steering wheel angle in corner entry
f	Rebound frequency
fa	Brake partition on the front axle
F	Generic force
Fc	Centrifugal force
Fg	Vertical force of the sprung mass of the generic semicorner on the single wheel at ground
Fs	Static axial force on the spring (spring preload)
$Flat$	Lateral force
Fyi	Lateral force on each axle that balances the yaw moment
G	Center of gravity

Analytical index

G_s	Center of gravity of the sprung mass
G,r	Yaw gain
γ	Camber angle
$h_{Rc,f}$	Front roll center height
$h_{Rc,r}$	Rear roll center height
H_G	Center of gravity height
h	Center of gravity height
I.C.	Instant center of the equivalent quadricycle
ICFV	Center of instant rotation of the suspension projected on the transverse plane
ICSV	Center of instant rotation of the suspension projected on the longitudinal plane
I_p	Polar moment of inertia of the circular section of the anti-roll bar
i,j,k	Versors of the axes x,y e z
J	Moment of inertia of the vehicle with respect to the z axis
φ	Inclination with respect to the horizontal plane of the joint between the Ct and the Rc
K_f	Roll stiffness of the front axle
K_b	Torsional stiffness of the anti-roll bar
$K_{bar,f}$	Stiffness in opposite wheel travel of the front axle due only to the bar
$K_{bar,r}$	Stiffness in opposite wheel travel of the rear axle due only to the bar
$K_{bar,susp}$	Stiffness of the bar on the attachment point at the suspension
KBETA	Sideslip angle gradient
K_{spring}	Spring stiffness
$K_{opp,f}$	Stiffness in opposite wheel travel of the front axle
$K_{opp,r}$	Stiffness in opposite wheel travel of the rear axle
K_r	Roll stiffness of the rear axle
KPI	King pin inclination
KPIoffset	Transversal arm at wheel center
K_t	Vertical stiffness of the tire
KROLL	Roll gradient
K_{susp}	Suspension stiffness at ground on the single wheel
$K_{susp,tire}$	Suspension stiffness at ground with tire, on the single wheel
K_{us}	Understeer gradient
$K_{us,vc}$	Understeer gradient at constant speed
K_x	Longitudinal stiffness of the bristle (brush element)
K_y	Transversal stiffness of the bristle (brush element)
L	Distance between the front axle and the intersection point of the steering leverages
L_b	Length of the anti-roll bar
L_c	Length of the anti-roll bar arms
L_{rel}	Relaxation length
L_{free}	Free length of the spring

Automotive suspension

L_{inst}	Installed length of the spring
L_{min}	Minimum length of the spring
l_a	Adhesion length of the contact patch
$Lambda$	Ratio between the displacement of the spring and the displacement of the wheel (at ground)
$Lambda,b$	Ratio between the displacement of the bar attachment and the displacement of the wheel (at ground)
$Lambda,_{bump}$	Ratio between bump stop clearance and bump stop clearance at ground
M	Total vehicle mass
M,yaw	Yaw moment
M_s	Sprung mass of the vehicle
M_{sem}	Sprung mass of the generic semicorner
M_{nsfl}	Front left unsprung mass
M_{nsrl}	Rear left unsprung mass
M_{srl}	Left rear sprung mass
$M,lift$	Lift moment
M_{ns}	Vehicle unsprung mass
$M_{ns,s}$	Unsprung mass of the generic semicorner
$M_{ns,f}$	Front unsprung mass
$M_{ns,r}$	Rear unsprung mass
$M,roll$	Roll moment
$\mu s,y$	Lateral static friction between the bristle and road
$\mu d,y$	Lateral dynamic friction between the bristle and road
μx	Coefficient of longitudinal adherence
μy	Coefficient of lateral adherence
p	Wheelbase
P_r	Peak of yaw rate in a steep steer
q	Corrective coefficient of the compression damping of the shock absorber
θ	Rotation vector around the instantaneous rotation axis of the suspension
θ	Torsion angle of the anti-roll bar
θ_m	Upright rotation angle around the steering axis
θ_{FT}	Equivalent front hinge inclination when braking (*Anti dive*)
$\theta_{0.7}$	Front braking force inclination with 70% distribution to the front
θ_{RT}	Equivalent rear hinge inclination when braking (*Anti lift*, rear)
$\theta_{0.3}$	Rear braking force inclination with 70% distribution to the front
θ_{FC}	Equivalent front hinge inclination in acceleration (*Anti lift*, front)
θ_1	Acceleration force inclination with 100% front or rear distribution
θ_{RC}	Equivalent rear hinge inclination in acceleration (*Anti squat*)
R_c	Roll center
R_c,f	Front roll center
R_c,r	Rear roll center

Analytical index

r	Yaw speed (yaw rate)
$r,stab$	Stabilized value of the yaw rate in a step steer
r,max	Maximum peak of yaw rate
$\dot{r}$	Yaw acceleration
R	Radius of curvature
Ra	Resistance to advancement by the tire
$RSCf$	Front tire under load radius
$RSCr$	Rear tire under load radius
SAO	Steering axis offset (Longitudinal arm at wheel center)
SC	Shear center of the central section of twist beam axle
Scf	Longitudinal displacement of the front wheel center
Scrub radius	Transversal arm at ground
Scr	Longitudinal displacement of the rear wheel center
Stf	Longitudinal displacement of the front contact patch center
Str	Longitudinal displacement of the rear contact patch center
SWA	Steering wheel angle
$SWA,0$	Steering wheel angle at zero speed
t	Wheel track at ground
t_a	Distribution of traction on the front axle
t_f	Front wheel track at ground
toe-in	Converging wheels/ closed wheels
toe-out	Diverging wheels/ opened wheels
t_r	Rear wheel track at ground
t_p	Distribution of traction on the rear axle
$TR\text{-}PP()$	Delta time between the peaks of the two quantities shown in brackets
τ	Steering ratio (steering wheel angle/steer angle)
τ_y	Elastic lateral stress on the bristle
Uy	Transversal displacement of the point on the road side of the bristle
$\mathbf{V}_C$	Wheel center velocity
$\mathbf{V}_G$	Velocity of center of gravity
Vx	Velocity of center of gravity along x
Vy	Velocity of center of gravity along y
Vr,max	Steering wheel angle velocity at maximum yaw rate peak
WC	Wheel center
ω	Tire angular velocity
$\mathbf{\Omega}$	Angular velocity of the vehicle in planar motion
Y	Centrifugal force
ψ	Roll angle
ζ	Longitudinal coordinate of the contact patch

Bibliography

Books

[1] M. Guiggiani, *Dinamica del veicolo*, Città Studi Edizioni, 2007.

[2] M. Guiggiani, *The science of vehicle dynamics*, Springer, 2014.

[3] H.B. Pacejka, *Tire and Vehicle Dynamics*, Elsevier, 2012.

[4] E. Zagatti, R. Zennaro, P. Pasqualetto, *L'assetto dell'autoveicolo*, Levrotto & Bella, 1998.

[5] W.F. Milliken, D.L. Milliken, *Race Car Vehicle Dynamics*, SAE, 1994.

[6] G. Della Valle, *Meccanica delle vibrazioni*, CUEN, 1988.

[7] F. Timpone, *Appunti del corso di meccanica del veicolo*, Università degli Studi di Napoli, Federico II.

[8] A. MORELLI, *Progetto dell'autoveicolo*, CELID, 2000.

[9] G.A. Pignone, U.R. Vercelli, *Motori ad alta potenza specifica*, Giorgio Nada Editore, 2003.

[10] T.D. Gillespie, *Fundamentals of Vehicle Dynamics*, SAE, 1992.

[11] J.C. DIXON, *Tires, Suspension and Handling*, SAE, 1996.

[12] G. Genta, L. Morello, *The automotive chassis*, Springer, 2014.

[13] *Racecar Engineering Magazine*, Giugno 2001, Luglio 2001, Agosto 2001.

Websites

www.racing-car-technology.com.au

www.racing-car-technology.com.au/Steering Ackerman4.doc

www.quattroruote.it

www.alfasport.net

www.autotecnica.org

www.evomagazine.it

Annotations

Automotive suspension

Annotations

Automotive suspension

Annotations